Hybrid Systems Based on Solid Oxide Fuel Cells

Hybrid Systems Based on Solid Oxide Fuel Cells

Modelling and Design

Mario L. Ferrari
University of Genova, Italy

Usman M. Damo
University of Manchester, UK

Ali Turan
University of Manchester, UK

David Sánchez
University of Seville, Spain

Registered Offices
John Wiley & Sons Inc., 111 River Street, Hoboken, NJ 07030, USA
John Wiley & Sons Ltd, The Atrium, Southern Gate, Chichester, West Sussex, PO19 8SQ, UK

Editorial Office
The Atrium, Southern Gate, Chichester, West Sussex, PO19 8SQ, UK

For details of our global editorial offices, customer services, and more information about Wiley products visit us at www.wiley.com.

Wiley also publishes its books in a variety of electronic formats and by print-on-demand. Some content that appears in standard print versions of this book may not be available in other formats.

Library of Congress Cataloging-in-Publication Data

Names: Ferrari, Mario L., 1978- author. | Damo, Usman M., 1984- author. |
    Turan, Ali, author. | Sánchez, David, 1977 April 14- author.
Title: Hybrid systems based on solid oxide fuel cells : modelling and design
    / Mario L. Ferrari, University of Genova, Italy, Usman M. Damo,
    University of Manchester, UK, Ali Turan, University of Manchester, UK,
    David Sánchez, University of Seville, Spain.
Description: First edition. | Hoboken, NJ, USA : John Wiley & Sons, Inc.,
    2017. | Includes bibliographical references and index.
Identifiers: LCCN 2017003602 (print) | LCCN 2017004069 (ebook) | ISBN
    9781119039051 (cloth) | ISBN 9781119039068 (pdf) | ISBN 9781119039075
    (epub)
Subjects: LCSH: Solid oxide fuel cells. | Hybrid power systems–Equipment and
    supplies. | Renewable energy sources.
Classification: LCC TK2933.S65 F47 2017 (print) | LCC TK2933.S65 (ebook) |
    DDC 621.31/2429–dc23
LC record available at https://lccn.loc.gov/2017003602

Cover Design: Wiley
Cover Images: (Bottom Image) © Fertnig/Gettyimages;
        (Top Image) Courtesy of the authors;
        (Background) © Max Krasnov/Shutterstock

Set in 10/12pt WarnockPro by SPi Global, Chennai, India
Printed and bound in Malaysia by Vivar Printing Sdn Bhd

10 9 8 7 6 5 4 3 2 1

Contents

Preface

Even though solid oxide fuel cell (SOFC) technology reached a significant development milestone around 30 years ago, no hybrid system prototypes were built before the 2000 Siemens-Westinghouse plant. Due to the enormous engineering system complexity and cost, SOFC/turbine hybrid plants only attracted substantial research interest at the end of the twentieth century when environmental concerns became very visible and demanding.

Considering the widespread enthusiasm regarding research and development activities for hybrid systems on the eve of the twenty-first century, a 'partial downsizing' is now apparent due to several unresolved engineering and sustainability problems and the ever-present, overriding cost and reliability issues. Thus, the forecast plans for commercialization carried out during the past decade seem to have failed to deliver acceptable hybrid system performance under realistic operational conditions, due to the various technological, complexity and cost issues.

Furthermore, comparing the developmental status of hybrid systems with state-of-the-art anticipated performance metrics of the past decade, several publications have now presented newly validated solutions for several of the previously outstanding applied research issues (such as cost decrease, SOFC/turbine coupling and control system development), incorporating significant promising technology improvements. Hence, it is essential to consider that concentrated funding resources are still necessary to profitably combat/resolve all of the technical issues and to reach the required high levels of reliability, high plant operative life and low-cost performance for acceptable commercial adoption on a wider scale. For such reasons, focused efforts and research interests/activities at both academic and industrial levels are absolutely essential. Even if hybrid systems will not be ready for commercialization in a few years, the extremely desirable performance and environmental aspects promised via this technology will be a central pillar for future energy generation and hydrogen economy development.

Due to the promising performance attributes and the recent substantial development of hybrid system technology based on solid oxide fuel cells (SOFCs), the authors have decided to develop this book to produce an updated text targeted at both practicing engineers and academic researchers. In comparison with previously published texts, the authors pay special attention to the latest research and development activities at both the theoretical and experimental levels. Thus, following the discussions of the basic aerothermodynamics and electrochemistry of the primary components (the SOFC stack and microturbine aspects are presented in Chapters 2 and 3), including updated descriptions covering the latest technological improvements and commercialization aspects, an innovative approach is considered to further develop the SOFC/turbine coupling in an individual chapter (Chapter 4). For that reason, special attention is devoted to system constraints, problem/solution details based on the latest academic/industrial research activities, and performance aspects of currently available commercial prototypes. Furthermore, the book presents details regarding hybrid system modelling activities from different points of view including theoretical/computational (Chapter 5) and physically based approaches (Chapter 6). In comparison with previous publications on SOFC based systems, this book devotes large sections and presents detailed discussion on experimental development devices collectively referred to as emulator rigs, as these tools are widely and routinely used to develop rational and profitable configurations covering hybrid systems based on SOFC and gas turbine systems. Currently, these experimental facilities show great potential regarding applied research for such power plants, and the results generated via their use are considered absolutely essential for solving several technical hardware and optimization issues for such hybrid systems.

Finally, various conflicting engineering issues and commercialization potentials to be pursued for the widespread adoption of such innovative and efficient power plants are discussed in Chapter 7, focusing special attention on future perspectives and possible solutions.

M.L. Ferrari
U.M. Damo
A. Turan
D. Sánchez

Acknowledgements

The authors would like to thank all the staff of the Thermo-chemical Power Group (TPG) of the University of Genoa for the shared experience involving theoretical and experimental activities and international collaboration opportunities. A special acknowledgement is due to Prof. Massardo Aristide F. (Director of the TPG) for his essential scientific support. The authors would also like to recognize and thank the fuel cell research group at the US Department of Energy, National Energy Technology Laboratory (NETL), Morgantown WV, US. To Dr Joseph Dawes is due a sincere note of thanks for the wonderful execution of the arduous task of going through the entire manuscript for both technical and language aspects of the material. Mr Ibrahim M. Damo deserves recognition for the redesign/reproduction of many figures in chapters 1 and 5. Also, the authors would like to thank Mr Che-Wei Nien, a graduate of the University of Manchester (MSc Thermal power), for his contribution to Chapter 5 with his thesis. Prof. Sánchez would like to gratefully acknowledge Gonzalo Sánchez-Martínez and José María Rodríguez at the University of Seville for their assistance in editing and largely improving the artwork in Chapter 3 on micro gas turbines.

Acknowledgements

1

Introduction

The current and future energy scenarios faced by the international community are discussed in this chapter, starting with a brief presentation of the energy landscape and related issues, including the increase in demand and environmental aspects. A list of possible solutions to existing and foreseen problems is presented and discussed, setting the framework to highlight the significant potential of fuel cells for future power generation. Following on from this, the performance characteristics of fuel cells are introduced, including an analysis of their different types and corresponding differential features. Additionally, attention is devoted to hybrid systems based on the coupling of high-temperature fuel cells and microturbines (mGTs).

1.1   World Population Growth, Energy Demand and its Future

A study carried out by the United States Census Bureau (USCB) [1] estimated that the world population exceeded 7 billion on

Hybrid Systems Based on Solid Oxide Fuel Cells: Modelling and Design, First Edition.
Mario L. Ferrari, Usman M. Damo, Ali Turan, and David Sánchez.
© 2017 John Wiley & Sons Ltd. Published 2017 by John Wiley & Sons Ltd.

12 March 2012. Now, at the time of writing in August 2016 with the global population standing at about 7.4 billion [2], this figure is expected to continue rising over the coming decades [2]. As the world population grows, in many countries faster than the global average of 2%, the need for more and more energy is intensifying in a somewhat similar proportion, thus putting pressure on the natural resources available and existing infrastructures. This higher energy consumption is not only due to the growth in world population, but also to the improved lifestyles leading to a greater energy demand per capita (two features that inevitably come together). This is best exemplified by the fact that the wealthy industrialized economies comprise 25% of the world's population but consume 75% of the world's energy supply [3]. A recent study (from ref. [4]) shows that the total world consumption of marketed energy is expected to increase from 549 quadrillion British thermal units (Btu) in 2012 to 629 quadrillion Btu in 2020, and to 815 quadrillion Btu in 2040 – a 48% increase from 2012 to 2040 [4].

Indeed, the landscape of future energy demand and generation projected for the world seems rather bleak, as most nations, including the most developed ones, depend primarily on conventional energy sources such as oil, coal and gas to generate power not only for the domestic and industrial sectors but also for transportation. This dependency results in global warming, contributes to rises in fuel prices that constitute a burden on economies, and can lead to delays in energy production and supply [5, 6]. Furthermore, even if the global production of fossil fuels is currently sufficient to cover the world's needs, the exponential rise in the exploitation rate of this finite, fast-depleting resource would pose a risk to the future energy demand and generation balance [7–9]. In the long run this global dependence on conventional fuel sources for power production will prove problematic because the world will eventually fall short or run out of these resources. Renewable energy sources are often set forth as a feasible alternative to this fossil-fuel dominated world [10], although many of their inherent features, such as their low energy density, intermittency and geographical distribution, pose a number of challenges that remain to be solved today.

1.2   World Energy Future

Due to the heavy reliance of most nations worldwide on fossil fuels for power generation and transportation, the atmospheric concentrations of carbon dioxide and methane have increased by 36% and 148% respectively, compared with pre-industrial levels [11]. These levels are indeed much higher than at any time during the last 800,000 years, the period for which reliable data have been extracted from ice cores. This observation is further confirmed by less direct geological observations that also show that carbon dioxide concentrations higher than today were last seen about 20 million years ago. These findings suggest that the root cause for such high concentrations is anthropogenic, mainly hydrocarbon-based fuel burning (responsible for three-quarters of the increase in CO_2 from human activity over the past 20 years) and deforestation [11]. Other environmental factors, including air pollution, acid precipitation, ozone depletion and emission of radioactive substances, are also of concern and raise awareness of the negative impact of human activity on the environment [3].

As a consequence of this massive production of anthropogenic carbon dioxide and other greenhouse gases (trace gases in particular [12]), global temperatures in 2016 were 0.87°C above the long-term 1880–2000 average (the 1880–2000 annually averaged combined land and ocean temperature is 13.9°C), which translates into a warming rate of around 0.61°C/century over the last few decades. In particular, the average temperature of the Atlantic, Pacific and Indian oceans (covering 72% of the Earth's surface) has risen by 0.06°C since 1995. The situation regarding global warming is far from being under control. As stated by the US Department of Energy's forecast, carbon emissions will increase by 54% above 1990 levels by 2015, making the Earth likely to warm by 1.7–4.9°C over the period 1990–2100 (see Figure 1.1). Such observations demonstrate the need for efforts towards alleviating energy-related environmental concerns in the near future [3].

Achieving higher efficiencies and, if possible, the utilization of renewable energies in power generation technologies will

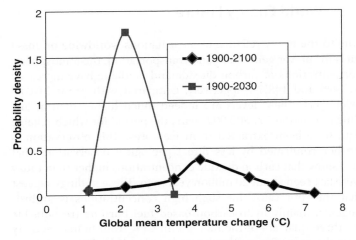

Figure 1.1 Global mean temperature probability changes, for the years 1990–2100 and 1990–2030. *Source*: Omer (2008) [3]. Reproduced with permission of Elsevier.

be vital steps in mitigating or reducing these environmental problems, whilst meeting the expected rise in energy demand in the future. With increasing fuel prices and significant pressure to reduce emissions, increasing energy efficiency is considered amongst the most feasible and cost-competitive approaches for reducing CO_2 emissions. For instance, Britain wastes 20% of its fossil fuel and electricity which, if used efficiently, would translate into a potential £10 billion annual reduction in the collective fuel bill and a reduction of some 120 million tonnes of CO_2 emissions [3]. Unfortunately, even if energy is currently recognized globally as being at the centre of the sustainable development paradigm, the industrial and social development paths favour energy consumption rather than conservation [3].

The significant fuel consumption and CO_2 emission issues have to link with the fact that conventional thermal power plants (regardless of the type of fuel used) cannot convert all of the thermal energy supply into useful (mechanical) work. In most cases, more than 50% of the heat added to the working cycle is rejected to the environment. Combined heat and power (CHP) installations are able to use a part of this heat, which would otherwise be wasted in a conventional power plant, to raise the overall first law efficiency to values higher than 80% for the best available technology [3]. This concept enables drastic reductions of the primary energy consumption and cost compared with the independent production of both forms of energy (electricity and thermal energy).

Complementary to energy conversion at high efficiency, substituting fossil fuels with renewable energy sources is envisaged as another means to tackle the aforecited social, economic and environmental problems. Renewable energies are broadly regarded as energy sources that are naturally replenished over a short timescale (i.e. in comparison to the lifetime of a human being), such as sunlight, wind, rain, tides, waves and geothermal heat. They have shown the potential to replace conventional fuels in various distinct areas, such as utility-scale electricity generation, hot water production/space heating, fuels for transportation, and rural (off-grid) energy services [13, 14]. Renewable energy sources have the potential to constitute the future energy sector's backbone, despite some evident shortcomings such as low density and inherent intermittency.

According to the REN21's 2014 report [15], renewables contributed 19% to the world's energy consumption in 2012, and 22% to electricity generation in 2013, using both traditional (biomass) and more innovative renewable energy technologies such as solar power, large wind farms and biofuels [10]. The importance of renewable energy sources has been disseminated widely, and several nations worldwide have decided to invest large sums of money in renewable technologies; such is the case in the US with a total investment of more than $214 billion in 2013, whereas other countries like China are following close behind [15].

Hybrid systems based on the coupling of a microturbine with a high-temperature fuel cell are highly regarded as a solution for future power generation due to their high efficiency, ultra-low emissions and their ability to run on fuels such as hydrogen produced from renewable sources. These systems can achieve very high efficiencies: more than 60% electrical efficiency using natural gas (depending on the low heating value). This efficiency is virtually independent of plant size due to the modular nature of these devices. Hybrid systems based on solid oxide fuel cells (SOFCs) are of particular interest, because they have the potential to overcome the main limitations of traditional power plants, and furthermore to meet the hurdles posed for the world's future energy need without worsening environmental issues.

1.3 Introduction to Fuel Cells and Associated Terms

A fuel cell is a device that converts the chemical energy in a fuel into electricity through an electrochemical reaction with oxygen. Hydrogen is commonly used in a fuel cell, but hydrocarbons such as natural gas and alcohols like methanol are also used. In contrast to batteries, fuel cells require a constant source of fuel and oxygen/air to sustain the chemical reaction and thus produce electricity as long as this input flow is supplied [16, 17].

A fuel cell typically consists of an anode (negative electrode), a cathode (positive electrode), and an electrolyte that allows charges to move between the two sides of the fuel cell [16, 17]. Direct current electricity is produced when electrons are drawn from the anode to the cathode. Typical layouts and choices of materials vary between fuel cell types. A classification of layouts/materials used in different types of fuel cells is shown in Table 1.1.

1.3.1 Background for Fuel Cells and Thermodynamic Principles

Fuel cells as power systems were first conceived and realized by Sir William Grove in 1839, using the experimental setup shown in Figure 1.2 [18]. His demonstration is the reverse of electrolysis: an electric current is produced through the process of combining hydrogen and oxygen to form water. This process is an 'electrochemical burning' reaction (although no real combustion is present in a fuel cell) which consumes hydrogen as fuel and produces electricity instead of heat.

Despite the fact that a large proportion of those with an interest in fuel cells are professionals with a background in heat engines, it is wise to revisit some fundamental concepts in chemical thermodynamics. To this aim, a brief introduction to the Gibbs potentials is provided below.

Table 1.1 Comparison of fuel cell technologies [17].

	PEFC	AFC	PAFC	MCFC	SOFC
Electrolyte	Hydrated polymeric ion exchange membranes	Mobilized or immobilized potassium hydroxide in an asbestos matrix	Immobilized liquid phosphoric acid in SiC	Immobilized liquid molten carbonate in LiAlO$_2$	Perovskites (ceramics)
Electrodes	Carbon	Transition metals	Carbon	Nickel and nickel oxide	Perovskite and perovskite/metal cermet
Catalyst	Platinum	Platinum	Platinum	Electrode material	Electrode material
Interconnect	Carbon or metal	Metal	Graphite	Stainless steel or nickel	Nickel, ceramic, or steel
Operating temperature	40–80°C	65–220°C	205°C	650°C	600–1000°C
Charge carrier	H$^+$	OH$^-$	H$^+$	CO$_3^{-2}$	O$^=$
External reformer for hydrocarbon fuels	Yes	Yes	Yes	No, for some fuels	No, for some fuels and cell designs
External shift conversion of CO to hydrogen	Yes, plus purification to remove trace CO	Yes, plus purification to remove trace CO and CO$_2$	Yes	No	No
Prime cell components	Carbon-based	Carbon-based	Graphite-based	Stainless-based	Ceramic
Product water management	Evaporative	Evaporative	Evaporative	Gaseous product	Gaseous product

Source: US Department of Energy.

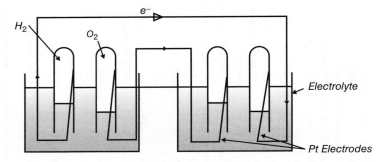

Figure 1.2 First demonstration of a fuel cell by Grove in 1839. *Source*: Srinivasan (2006) [18]. Reproduced with permission of Springer.

Let Γ be a chemical system whose absolute pressure, temperature, volume and entropy are denoted by p, T, v and S respectively. The following four Gibbs potentials are defined to calculate the work yielded by the system: internal energy (U), Helmholtz free energy (A), enthalpy (H), and Gibbs free energy (G), as shown in equations 1.1–1.3.

$$A = U - TS \qquad (1.1)$$

$$H = U + pV \qquad (1.2)$$

$$G = H - TS \qquad (1.3)$$

These Gibbs potentials are used in the analysis of a wide range of processes.

The total work yielded by the system Γ during an infinitesimal process, W, is higher than the amount of work that can actually be employed by a potential user, W_u [19]. The difference between both works is that done by the system on its surroundings:

$$dW = dW_u + d(pv) \qquad (1.4)$$

Based on this consideration, the four state functions previously listed are used to equate the First and Second Laws of thermodynamics applied to a closed system (Eqs 1.5 and 1.6), where work W and heat Q are considered positive when they flow out of and into the system respectively:

$$dU = dQ - dW \qquad (1.5)$$

$$dS = dQ/T + dS' \qquad (1.6)$$

Combining equations 1.2, 1.4 and 1.5 yields the following alternative form of the First Law:

$$dH = dQ - dW_u \tag{1.7}$$

Equations 1.5 and 1.6 can be interpreted in the following terms:

- If the process followed by the system Γ does not perform work, internal energy change equals heat added to the system ($dU = dQ$).
- If the process followed by the system Γ does not perform useful work, enthalpy change equals heat added to the system ($dH = dQ$).
- If the system undergoes an adiabatic process:
  - Total work equals internal energy decrease ($dW = -dU$).
  - Total useful work equals enthalpy decrease ($dW_u = -dH$).

Equation 1.6 evidences that the entropy gain of the system dS comes about due to the heat added to the system at constant temperature T plus a certain amount of unbalanced entropy change dS'. As stated originally by Clausius, this unbalanced entropy change is either positive in the case of an irreversible process, or null in the case of a reversible process [20]. Again, the combination of equations 1.1–1.3 and the First and Second Laws provide the following useful interpretations:

- If the system undergoes an isentropic process:
  - The total work done by the system equals the internal energy drop minus the energy dissipated to the surroundings ($dW = -dU - TdS'$).
  - The useful work done by the system equals the enthalpy drop minus the energy dissipated to the surroundings ($dW_u = -dH - TdS'$).
- If the system undergoes an isothermal process:
  - The total work done by the system equals the Helmholtz free energy drop minus the energy dissipated to the surroundings ($dW = -dA - TdS'$).
  - The useful work done by the system equals the Gibbs free energy drop minus the energy dissipated to the surroundings ($dW_u = -dH - TdS'$).

This set of useful thermodynamic relations provides a means to calculate the total and useful work through the four Gibbs potentials in a variety of processes. These relations are summarized in Table 1.2 for clarity.

Despite a thermodynamic analysis utilizing the considerations above showing the promise of the experimental setup shown in Figure 1.2, the current generated is usually very small due to the unfavourable characteristics of the three-phase interface (contact area between the gas, the electrolyte, and the electrode) and the high ion-transport resistance of the electrolyte [21]. These issues have driven the development of fuel cell technology towards flat and porous electrodes with a thin layer of electrolyte, as shown in Figure 1.3. Layouts of this kind result in the contact area being maximized and the resistance kept to a minimum, increasing the current produced [21].

Table 1.2 Useful relations based on work and heat potentials.

Process	−dU	−dA	−dH	−dG
$dW = 0$	$-dQ$			
$dW_u = 0$			$-dQ$	
$dQ = 0$	dW		dW_u	
$dS = 0$	dW_{max}		$dW_{u,max}$	
$dT = 0$		dW_{max}		$dW_{u,max}$

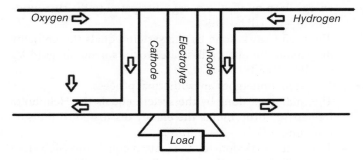

Figure 1.3 Basic fuel cell arrangement. *Source*: Larminie *et al.* (2003) [21]. Reproduced with permission from John Wiley & Sons.

1.3.2 Solid Oxide Fuel Cells (SOFCs)

Solid oxide fuel cells (SOFCs), as the name implies, are completely solid-state entities that use ceramic electrolytes. The development of SOFCs can be traced back to the 1890s when Nernst discovered that stabilized zirconia (ZrO_2) could conduct ions at certain temperatures, making zirconia a potentially useful electrolyte [21, 22]. Major manufacturers and their development focuses are listed in Table 1.3. A further investigation was carried out by Baur and Preis in 1943, showing that zirconia could serve as an oxygen-ion conducting electrolyte in fuel cells [22].

Today, this type of fuel cell is a high-temperature, solid-state electrochemical conversion device that produces electricity directly from electrochemical (oxidation) reactions. The cell operates at 600–1000°C where ionic conduction of oxygen ions takes place. Commonly, the anode is a Ni–ZrO_2 cermet and the cathode is Sr-doped $LaMnO_3$. Not using a liquid electrolyte avoids the attendant material corrosion and/or electrolyte management issues. The high temperature of the SOFC, nonetheless, poses stringent requirements on its materials, and hence the development of low-cost materials and the low-cost fabrication of ceramic structures that still fulfil the technical requirements are now the key technical challenges for the future utilization of SOFCs [17].

The basic operation of a SOFC is sketched in Figure 1.4. Oxygen (O_2) is reduced at the cathode–electrolyte interface forming oxygen ions ($O^=$) that are transported to the electrode through the electrolyte. Once at the interface between anode and electrolyte, these react with the hydrogen ions (H^+) to form water (H_2O) that is disposed of via the exhaust stream. Electrons are released at the anode and flow through the external load to the cathode, where they are used to reduce the oxygen molecules. This operating principle is summarized in Figure 1.4 and through the following reactions [23–25] (Eqs 1.8–1.10):

$$\text{Cathode}: \quad O_2 + 4e^- \rightarrow 2O^{2-} \qquad (1.8)$$

$$\text{Anode}: \quad 2H_2 + 2O^{2-} \rightarrow 2H_2O + 4e^- \qquad (1.9)$$

$$2CO + 2O^{2-} \rightarrow 2CO_2 + 4e^- \qquad (1.10)$$

Owing to the overriding influence of the reaction shown in Equation 1.9, the reaction represented by Equation 1.10 is not

Table 1.3 Major manufacturers of fuel cells.

Manufacturer	Developments
AFC Energy, Cranleigh, Surrey, United Kingdom (2006).	Alkaline fuel cells.
Apollo Energy Systems, Pompano Beach, Florida, USA (1966).	Develops, produces and markets fuel cell power plants, electrical propulsion systems, and alternative energy generation equipment.
Ballard Power Systems, Burnaby, British Columbia, Canada (1979).	Designs, develops, and manufactures zero-emission proton-exchange-membrane fuel cells. Ballard Power Systems, Inc. is a global leader in PEM (proton exchange membrane) fuel cell technology.
Doosan Fuel Cell America, Sunnyvale, California, USA (HQ) (2003).	A fuel cell manufacturer focusing on the stationary fuel cell and small business markets.
Intelligent Energy, Loughborough, United Kingdom (2001).	Specialists in the development of proton exchange membrane (PEM) fuel cells for application in the automotive, consumer electronics and stationary power markets.
UTC Power, South Windsor, Connecticut (1958).	Produces/develops fuel cells for numerous applications including for use in space and submarines.
SOFCpower SpA 2006, Trentino, Italy.	For stationary applications with electrical power requirements below 6 kW.
Mitsubishi Heavy Industries (MHI), established in 1914, Tokyo.	First domestic operation of a combined-cycle system combining SOFC and a micro gas turbine. Maximum power output of up 200 from 21 kW.
Rolls-Royce Fuel Cell Systems (RRFCS), 2002, Loughborough, UK.	Stationary power generation, applications range from 250 kW to >1 MW.
Hexis AG, 1997, Switzerland.	For stationary applications with electrical power requirements below 10 kW.
Siemens-Westinghouse, in SOFC/mGT business for more than 3 decades, USA.	For stationary applications with electrical power requirements above 100 kW.

Figure 1.4 Basic SOFC operation.

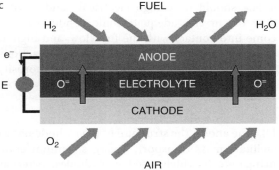

further considered in the analysis regarding the electrochemical reaction set. Thus the overall electrochemical reaction (Eq. 1.11) is obtained by adding equations 1.8 and 1.9:

$$\left.\begin{array}{l} \Rightarrow \quad 2H_2 + O_2 \rightarrow 2H_2O \\ \Rightarrow \quad H_2 + \frac{1}{2}O_2 \rightarrow H_2O \end{array}\right\} \tag{1.11}$$

1.3.2.1 Electrolyte

Currently, the majority of SOFC developers use electrolytes made of zirconia stabilized by a small amount of yttria (3, 8, or 10%), namely YSZ [17]. When the temperature is raised to more than 800°C, such electrolytes become good conductors of oxygen ions and show minimal electrical conductivity (i.e. transportation of electrons) [17–22]. As a main shortcoming, the very high operating temperature of the cell makes material selection very difficult, although the high temperature also provides an exploitable characteristic for hybrid systems.

1.3.2.2 Anode

Present SOFCs use anodes made from zirconia cermet (a mixture of ceramic and metal) [17, 21]. The metal used is nickel, chosen primarily due to its high electrical conductivity and stability under chemically reducing conditions [17, 21]. Nickel is preferred to platinum, the metal of choice in low-temperature fuel cells, due to its much lower cost, with only a small trade-off in transport properties. The zirconia used is both to inhibit the sintering of the metal particles and to provide a thermal expansion ratio close to that of the electrolyte. A well-designed anode should have high electrical conductivity to allow the flow of current, adequate ionic conductivity such that ions come into contact with the fuel flow, and high activity for the electrochemical and fuel conditioning[1] reactions [17]. At the same time, the anode should have a high porosity (20–40%) to allow mass transport of gases [17, 21].

1.3.2.3 Cathode

Like the anode, the structure of the cathode has to be porous to facilitate mass transport. Today, strontium-doped lanthanum manganite is widely used in cathodes. Alternative materials include P-type conducting perovskite structures that display mixed ionic and electrical conductivity; these are specifically worthy of consideration for lower temperature operation (about 650°C) as voltage loss becomes significant for the other standard materials [17, 21].

1.3.2.4 Interconnector

An interconnector exists to provide physical connection between neighbouring fuel cells so that a higher current can be produced by operating several fuel cells assembled in parallel. Under normal circumstances the only requirement for the interconnector material is that it has to have very high electrical conductivity. However, complications arise as SOFCs have a very high operating temperature. Problems such as different thermal expansion coefficients, cathode poisoning, and oxidation of metal become serious. For these reasons, at present, ceramic material (lanthanum chromite) is the favoured choice for tubular design SOFCs [17].

The components can be assembled together in various configurations (tubular, planar, etc.) each one of which exhibits advantages and disadvantages. As an example, Figure 1.5 illustrates a conventional cathode-supported tubular layout, which was dominant in SOFC technology in the late 1990s and early 2000s.

---

1 Fuel conditioning reactions refer to the reforming of natural gas and the water–gas shift, producing hydrogen from natural gas and carbon monoxide respectively. These two reactions produce the hydrogen that is eventually oxidized by the fuel cell.

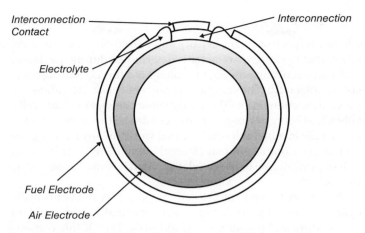

Interconnection
Contact

Interconnection

Electrolyte

Fuel Electrode

Air Electrode

Figure 1.5 Tubular SOFC module configuration [17]. *Source*: US Department of Energy.

1.3.3 Fuel Cell Reactions

The operation of a fuel cell implies a number of electrochemical reactions employing hydrogen and oxygen as fuel and oxidant to generate power (even if some fuel cells are able to operate on different fuels, the main fuel for these reactors is still hydrogen). However, unlike oxygen, which can be easily obtained from air, hydrogen is not found naturally and it must be produced from other hydrogen-containing compounds, typically hydrocarbons and/or water. There are different ways to extract hydrogen from conventional hydrocarbon fuels, amongst which the normal industrial practice is to use fuel reforming [21, 26, 27]:

1) Water electrolysis.
2) Steam reforming.
3) Partial oxidation.
4) Auto thermal reforming.
5) Coal gasification.

1.3.4 Fuel Cell Performance

The performance of a fuel cell is often related to fuel cell voltage directly:

$$P = V \cdot I \tag{1.12}$$

$$P = V \cdot J \cdot A \tag{1.13}$$

where J refers to current density and A to fuel cell active area. Given that the latter parameter is determined during the design and manufacturing process, the voltage of a fuel cell during operation is affected by the variations in current density. Recalling the previous section on the thermodynamic principles of fuel cells, Gibbs free energy changes are used to define the work potential of the cell. In a fuel cell, the 'external work' involves driving the flow of electrons around an external circuit. Any work done by a change in volume between inlet and outlet is not harnessed by the fuel cell.

The reference point of zero energy in a fuel cell is normally defined as that of pure elements, in the normal state at standard temperature and pressure (25°C, 0.1 MPa) [21]. If this convention is adopted, then the term 'Gibbs free energy of formation', G_f, rather than the 'Gibbs free energy', is used (akin to the terms 'enthalpy of formation' rather than just 'enthalpy' [21]). For an ordinary hydrogen fuel cell operating at standard temperature and pressure (STP), this means that the Gibbs free energy of formation of the input is zero, which is a useful simplification [21]. This simplification (or standardization) is possible because it is the change in energy that is important, meaning that it is the change in this Gibbs free energy of formation, ΔG_f, that gives the work potential of the fuel cell, that is, the difference between the Gibbs free energy of the products and the Gibbs free energy of the inputs or reactants [21]:

$$\Delta G_f = G_f \text{ of products } - G_f \text{ of reactants}$$

To make comparisons easier, it is convenient to consider these quantities in their molar-specific form. This indicated by a line over the lower-case letter, for example, $\overline{g}_{f,H_2O}$ is the molar-specific Gibbs free energy of formation for water [21].

The Gibbs potential is defined as:

$$G = H - TS \tag{1.14}$$

which is equivalent to:

$$\overline{g}_f = \overline{h}_f - T\overline{s} \tag{1.15}$$

For fuel cells, it is the change in Gibbs free energy that is responsible for the voltage induced. Therefore the following holds true:

$$\Delta \overline{g}_f = \Delta \overline{h}_f - T\Delta \overline{s} \tag{1.16}$$

Considering Equation 1.11, it can be observed that for every mole of hydrogen consumed (H_2), one mole of oxygen ions is used ($O^=$); one mole of water (H_2O) and two moles of electrons (e^-) are produced. It thus follows that the amount of electrons produced is then $2 \cdot N$, N being the moles of hydrogen consumed; this is known as Faraday's law. Assuming one electron carries with it $-e$ charge, the charge flow is $-2 \cdot N \cdot (-e)$ which equals to $-2F$, where F is Faraday's constant which is defined as the electric charge of a mole of electrons.

Upon development of the electrochemical reactions, a voltage difference between electrodes is built up in the cell, which drives the flow of electrons from the cathode to the anode. On the assumption that there are no losses (the ideal case), this is called the Nernst potential and is denoted by E. The electrical work (joules) developed by an ideal fuel cell is then:

$$W_{el} = E \cdot I \cdot t = 2 \cdot F \cdot E \tag{1.17}$$

where $I \cdot t$ is the flow of electrical charge passing through during a period of time t. This equation can now be linked to the change in Gibbs free energy through the information in Table 1.2:

$$\overline{g}_f = 2 \cdot F \cdot E \tag{1.18}$$

$$E = \frac{\overline{g}_f}{2 \cdot F} \tag{1.19}$$

This ideal potential of a cell operating at constant temperature (E), which yields the maximum electrical work output that the cell can produce, is not attainable in practice. Indeed, the kinetics of the reactions, their activation energy, the limited mass diffusion rate of reactants (to get to the reaction sites) and products (to be evacuated from the reaction sites), and the limited electronic/ionic conductivity of the fuel cell constituents, introduce inefficiencies that manifest as voltage losses. These losses are explained in more detail later in this chapter. The impact of pressure and temperature without these losses (i.e. on the ideal voltage of a fuel cell) are discussion in the next section.

1.3.5 Pressure and Concentration Effects

The foregoing discussion about the ideal cell potential E assumed the selection of a reference pressure and temperature (in this case 25°C and 1 atm), thus utilizing the standard cell potential E^0. However, in order to reduce the ionic resistivity of the electrolyte when integrated into hybrid systems, solid oxide fuel cells operate at a much higher temperature, and also typically operate at a higher pressure. This higher operating temperature and pressure means that the above expressions must be applied to Gibbs free energy changes at higher pressure and temperature or, more commonly, a correction must be applied to account for the effects of these two thermodynamic variables and of the concentration of reactants and products (whose proportion is not stoichiometric in a practical case). This correction is a mathematical expression of Le Chatelier's principle.

The correction for temperature can be introduced directly into the standard cell voltage E^0, which is evaluated at the operating temperature of the fuel cell. The corrections for pressure and composition are expressed as a complement to this. The resulting cell voltage is known as Nernst's potential and stands for the actual work potential of the fuel cell; that is, the voltage of a fuel cell operating at a given pressure, temperature and gas composition. A detailed derivation of the equation is not described here, but can be found in books and open literature [17, 21].

Considering the general reaction:

$$jJ + kK \leftrightarrow mM \tag{1.20}$$

The activities of the reactants and products modify the Gibbs free energy change of the reaction. Using thermodynamic arguments [28], it can be shown that the general Nernst's equation related to the reaction can be expressed as:

$$\bar{g}_f = \bar{g}_f^0 + R \cdot T \ln \frac{P_J^j P_K^k}{P_M^m} \tag{1.21}$$

where $\bar{g}_f^0$ is the change in molar Gibbs free energy of formation at standard pressure and the operating temperature of the fuel cell. For the particular case of the hydrogen oxidation reaction, this equation has the following aspect:

$$\bar{g}_f = \bar{g}_f^0 + R \cdot T \ln \frac{P_{H_2} P_{O_2}^{1/2}}{P_{H_2O}} \tag{1.22}$$

The impact of the activities of reactants and products on the Gibbs free energy change is clear in Equation 1.22. The impact on Nernst's potential, and thus on electrical work, is analogous and can be easily assessed if Equation 1.21 is used, obtaining:

$$E = \frac{\bar{g}_f}{2 \cdot F} + \frac{R \cdot T}{2 \cdot F} \ln \frac{P_{H_2} P_{O_2}^{1/2}}{P_{H_2O}} = E^0 + \frac{R \cdot T}{2 \cdot F} \ln \frac{P_{H_2} P_{O_2}^{1/2}}{P_{H_2O}} \tag{1.23}$$

where the partial pressure of each species is calculated through their corresponding molar fraction according to Dalton's law. If α, β and γ are the molar fractions of hydrogen and steam in the anode and oxygen in the cathode, and P_{an} and P_{ca} are the absolute pressures of the anode and cathode respectively:

$$P_{H_2} = \alpha \cdot P_{an}; \qquad P_{H_2O} = \beta \cdot P_{an}; \qquad P_{O_2} = \gamma \cdot P_{ca} \tag{1.24}$$

These expressions can be introduced into Equation 1.23 to calculate Nernst's potential. With the assumption that the absolute pressures in the cathode and anode are the same (which is a reasonable assumption based on mechanical considerations), this yields to:

$$E = E^0 + \frac{R \cdot T}{2 \cdot F} \ln \left(\frac{\alpha \cdot \beta^{1/2}}{\gamma} P_{cell}^{1/2} \right) \tag{1.25}$$

$$E = E^0 + \frac{R \cdot T}{2 \cdot F} \ln \left(\frac{\alpha \cdot \beta^{1/2}}{\gamma} \right) + \frac{R \cdot T}{4 \cdot F} \ln(P_{cell}) \tag{1.26}$$

where the effects of temperature (through E^0), composition and pressure on Nernst's potential can clearly be seen.

1.3.6 Irreversibilities in Fuel Cells

The performance of a SOFC operating at 800°C under atmospheric pressure is shown in Figure 1.6. The ideal Nernst's

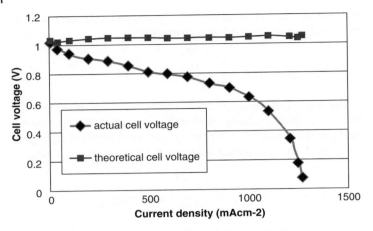

Figure 1.6 SOFC performance at 800°C under atmospheric pressure (graph generated using expression provided in [17]). *Source*: US Department of Energy.

potential is shown by the square symbols and associated line, whilst the diamond shapes represent the actual voltage differences between electrodes. It is observed that the latter is substantially lower than the theoretical limit, indicating that fundamental losses arise under normal operating conditions. As plotted in Figure 1.6, there is a rapid initial fall of cell voltage, followed by a slow and linear decrease. A final sudden decrease is also present at high current density.

There are several reasons for this voltage drop with respect to the ideal value. The four most important are: activation losses; fuel crossover and internal currents; ohmic losses; and mass diffusion or concentration loss. The effects of the four types of losses are often easy to distinguish in a V–I diagram, especially for low-temperature fuel cells, as illustrated in Figure 1.7 [17]. A brief discussion of the relevant losses is given in the following paragraphs, while interested readers are directed to accessible government-contracted reports and other documents for more details on how these losses are described numerically [17, 21].

- *Activation-related losses*: these stem from the activation energy of the electrochemical reactions at the electrodes. These losses are primarily due to the sluggishness of the reactions at hand, the electrocatalyst material and microstructure, reactant activities (and hence utilization), and are influenced very weakly by current density [17, 21].

- *Fuel crossover and internal currents*: these arise as fuel and electrons migrate through the electrolyte, due the non-ideal nature of this element. For example, in a hydrogen fuel cell, electrons proceeding through the electrolyte instead of through the external circuit do not contribute to the production of electrical work.
- *Ohmic losses*: ohmic losses are brought about by ionic resistances in the electrolyte and electrodes, electronic resistance in the electrodes, current collectors and interconnects, and contact resistances. Ohmic losses are proportional to the current density, and depend upon the material selection, stack geometry and operating temperature [17, 21].
- *Mass-transport-related losses*: these are due to the limited mass diffusion rates of reactants and products at high current density conditions [17, 21].

An overview of the physics behind each loss, further to the descriptions above, is now provided. Reactions occurring on the electrodes are not instantaneous and require both a finite time to proceed and an initial threshold voltage to be met in order to overcome the activation energy of the reaction. Losses associated with these factors are known as activation

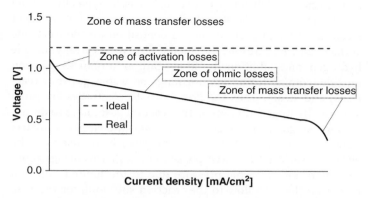

Figure 1.7 Ideal and actual fuel cell voltage/current characteristics [17]. *Source*: US Department of Energy.

losses and, for the cited reasons, the current density has little to no influence on the size of these losses. The electrolyte in fuel cells is designed to allow the flow of ions only [28, 29]. Nevertheless, in practice, due to the non-ideal nature of this element, fuel crossover and internal currents arise as fuel and electrons migrate through the electrolyte. For example, in a hydrogen fuel cell, electrons proceeding through the electrolyte instead of through the external circuit do not contribute to the production of electrical work. Moreover, according to Faraday's law, when a hydrogen molecule flows through the electrolyte, two moles of electrons are wasted corresponding to the anodic half-reaction (Equation 1.7) that fails to take place. Of all the losses, ohmic losses are the easiest to comprehend. All physical components of a fuel cell have finite conductivities, meaning each pose certain conductive resistances impeding the free flow of electrical charge carriers. Therefore, when ionic flow passes through the electrolyte or when electrons migrate past the electrodes and interconnectors, linear voltage drops occur according to Ohm's law.

As previously mentioned, fuel cells need oxygen and hydrogen as reactants to produce electrical current. The primary source of oxygen is air, resulting in the normal practice of adopting atmospheric air to provide the oxidant. This air stream is typically inducted to the cathode [28]. As previously discussed, hydrogen is generally produced through fuel reforming which, in the case of high-temperature fuel cells, is typically carried out in the anode (at least partially) in the presence of sufficient steam in order to prevent carbon deposition on the electrode surface [28]. As a result, the anodic flow will typically include hydrogen and additionally some carbon monoxide, carbon dioxide (coming from the reforming and water–gas shift reactions) and steam. Thus the flows at both electrodes are often not pure hydrogen or oxygen. The reactant gases have to diffuse through stagnant non-reacting gases and the electrode matrix structures to replenish the reactant gases being used on the electrode surface (the three-phase interface). This brings about regions of lower partial pressure due to the limited diffusion rate from the core flow to the reaction sites and, for the case of steam, vice versa. These lower partial pressures of reactants and higher partial pressure of products (where not adequately evacuated from the reaction sites) cause a voltage drop as deduced from Equation 1.23. This is termed mass transport loss or concentration loss [17, 18].

1.3.7 Fuel Cell Applications

Fuel cell technologies can be grouped into three primary areas: portable power generation, stationary power generation, and power for transportation. Further categorizations may be made according to fuel type and associated infrastructure. These categorizations are particularly important relating to the production, distribution, storage and dispensing of the fuels, crucial factors in implementing fuel cell technology. As a result of the inherent size flexibility of fuel cells, the technology may be used in applications with a broad range of power and plant size requirements. Fuel cells have application ranges from systems of a few watts to megawatts. Mobile fuel cell applications primarily include transportation systems and portable electronic equipment, while stationary applications primarily include combined heat and power systems for both residential and commercial needs [30, 31].

1.3.7.1 Transportation Applications

Nearly all leading car manufacturers have designed and developed at least one prototype vehicle using fuel cells [30, 31]. Manufacturers such as Toyota and Ford have chosen to feed the fuel cell with methanol, whilst others have used pure hydrogen – Opel using liquid hydrogen and General Motors storing hydrogen in hydride metals. There is a general short-term trend for auto manufacturers to use fuel cells using reformed methanol. However, as evidenced in the open literature, hydrogen still remains the fuel of choice in fuel cells for the majority of car manufacturers, mainly due to the inherent simplification of the fuel processing stage. The auto manufacturers' interest in fuel cells is not a new development. Since 1994, Daimler-Benz working in agreement with Ballard built a series of PEMFC powered cars, the first of which was fuelled with hydrogen, and in 1997 Daimler-Benz released a methanol-fuelled car with a 640 km range [30].

Furthermore, in 1993, Ballard Power Systems demonstrated a 10-metre light-duty transit bus with a 120 kW fuel cell system, followed by a 200 kW, 12-metre heavy-duty transit bus in 1995. These buses use no traction batteries and operate using compressed hydrogen as the onboard fuel (which is a similar technology to that used in buses running on compressed natural gas stored over the roof in high pressure cylinders) [30].

1.3.7.2 Portable Electronic Equipment

In addition to large-scale power production, miniature fuel cells can be profitably/potentially employed as batteries to power consumer electronic products such as cellular telephones, portable computers, and video cameras [30, 31]. Small fuel cells could also be used to power telecommunications satellites, augmenting solar panel performance, whereas micro-machined fuel cells could provide power to computer chips. Finally, micro/nano fuel cells could safely produce power for biological applications, such as hearing aids and pacemakers [30]. Contrary to transportation applications where much development has been required to bring systems incorporating fuel cells to the level of maturity where they can provide a credible alternative to internal combustion engines in producing a mechanical output, in portable electronic equipment, fuel cells compete with devices such as batteries in directly producing an electrical output [30].

1.4 Gas Turbines

Gas turbines (normally microturbines) are essential components to obtain high efficiency in hybrid systems. Their main components include an upstream compressor coupled to a downstream turbine, with a combustion chamber lying in-between. In small devices called microturbines, heat exchangers are also common in order to reduce fuel consumption. Finally, an electric generator converts the shaft work into electricity. A picture of the Turbec T100 microturbine (100 kWe) is shown in Figure 1.8.

1.4.1 Background of Gas Turbines

The gas turbine was hailed as one of the greatest discoveries of the twentieth century and has, since the 1940s, been used for numerous applications from power generation to

Figure 1.8 A Turbec T100 microturbine.

aircraft propulsion [32] due to its high reliability, efficiency, and cost-effectiveness. Further details related to gas turbines, specifically small-sized machines (called microturbines) based on the recuperated Brayton cycle and vitally important in hybrid system development, may be found in Chapter 3.

1.5 Coupling of Microturbines with Fuel Cells to Obtain 'Hybrid Systems'

The integration of a microturbine with a fuel cell to provide a power generation system with very high efficiency is generally termed as 'hybrid'. Combining the two distinct systems increases the efficiency with respect to using a microturbine alone, from values between 25% and 30% to values close to 65% (low heating value (LHV) fuel to electricity). This increase in efficiency is due to the contribution of the fuel cell to power generation which is based on electrochemical energy conversion and thus not subjected to the limitations of the Carnot cycle. In a typical high-temperature hybrid system based on a SOFC and microturbine, such as those in [32–35], the air stream is first pressurized through a compressor and then supplied to the SOFC where fuel is added and the electrochemical conversion of chemical energy into electrical work takes place. The high-temperature exhaust stream from the fuel cell is then discharged into an expander which provides work to drive the compressor and electric generator. The operation of the cell at high pressure (corresponding to a higher voltage) and the additional work produced by the gas turbine are significant upgrades to a standard atmospheric pressure fuel cell. This twofold enhancement significantly raises system efficiency [32].

On the negative side, these integrated systems (fuel cells and micro gas turbines) have a number of critical design constraints/limitations that require serious operational and developmental consideration, both in stationary and transient operation, which will be explored in later chapters.

Examples of companies currently engaged in the development of hybrid system prototypes are Siemens-Westinghouse, Rolls-Royce Fuel Cell Systems and Mitsubishi Heavy Industries. Westinghouse has been engaged in SOFC development for more than three decades. Following a merger in 1998 the company later became Siemens-Westinghouse Power Corporation (SWPC), focusing on the research and development (R&D) of SOFC-mGT hybrid systems for the emerging distributed power market. This company developed a full scale 100 kW plant (without an integrated micro gas turbine) which operated for more than 15,000 hours. This plant met many of the testing targets, although the measured performance values were significantly lower than the potential targets of these plants [33]. Other improved plants over the years have been designed, developed and tested, as mentioned in the rest of the book.

Table 1.4 Overview of some groups performing research in SOFC and SOFC-mGT hybrid systems [32, 36].

Institution	Location	Main staff	Models	Current focuses
NETL/D.O.E.	Morgantown, USA	Tucker, D. Shelton, M.	Transient model	Dynamics, control, start-up emulator
Siemens-Westinghouse	Erlangen, Germany	Hussmann, K. Ciesar, J.A.	Tubular SOFCs	Part-load, dynamics, control strategy, cycle layout and design
Rolls-Royce Fuel Cell Systems	Derby, UK	Agnew, G. Bozzolo, M.	IP-SOFC (RR) Hybrid systems	Part-load, dynamics, control strategy, cycle layout and design
German Aerospace Center (DLR), Institute of Combustion Technology	Stuttgart, Germany	Hohloch, M. Widenhorn, A. Lebküchner, D. Panne, T. Aigner, M.	Tubular SOFCs	Cycle layout and design emulators
Advanced Power and Energy Program (APEP) at the University of California, Irvine	California, Irvine, USA	Samuelsen, S. Brouwer, J. Rao, A.D. Yi, Y.	Tubular SOFCs	Part-load, dynamics, control strategy, cycle layout and design
TPG	Genova, Italy	Massardo, A.F. Costamagna, P. Magistri, L. Traverso, A. Ferrari, M.L.	Tubular SOFCs IP-SOFC (RR) Hybrid systems	Part-load, dynamics, control strategy, cycle layout and design

Table 1.4 (Continued)

Institution	Location	Main staff	Models	Current focuses
Politecnico di Milano	Milano, Italy	Campanari, S. Iora, P.	Tubular SOFCs	Cycle design, part-load
VOK-LTH	Lund, Sweden	Assadi, M. Torisson, T. Selimovic, A. Kemm, M. Hildebrandt, A.	Planar SOFC, detailed compressor	Dynamics, control, compressor surge modelling
SNU	Seoul, Korea	Song, T.W. Sohn, J.L.	Tubular SOFCs	Cycle layout and design
NTNU-EPT	Trondheim, Norway	Bolland, O. Thorud, B. Stiller, C.	Tubular SOFCs	Part-load, dynamics, control
NTNU-ITK	Trondheim, Norway	Imsland, L. Kandepu, R.	Lumped model	Dynamics, control
School of Automation, University of Electronic Science and Technology of China and Institute of Fuel Cell, Shanghai Jiao Tong University, Shanghai, China	China	Wua, X. Zhub, X.	Hybrid systems	Part-load, dynamics, control strategy, cycle layout and design

Source: Adapted from Damo (2016) and Stiller (2006).

1.5.1 Active Hybrid Systems Research Groups

Due the importance of the related research activities in this field, Table 1.4 lists some research groups that are actively working in the field of hybrid system modelling and contributing to this field of knowledge in the form of publications based on experimental activities and/or theoretical development/analysis [32, 36].

1.6 Conclusions

Hybrid systems are expected to play a large role in the improvement of power generation systems. Manufacturers and researchers are devoting special efforts to developing this technology and bridging the gap to the market. However, at the moment these systems are not ready for commercialization due to both technical problems that are yet to be adequately solved, and the high cost of materials used to produce the fuel cells and related components. Hence, research and development activities including theoretical investigations and, primarily, experimental work are of critical importance.

References

1 US Census Bureau (2016) *World Population Clock*. Available at https://www.census.gov/popclock/
2 United Nations Population Fund (2015) *The State of World Population*. Available at http://www.unfpa.org
3 Omer, A.M. (2008) Energy, environment and sustainable development. *Renewable and Sustainable Energy Reviews*, 12 (9), 2265–2300.
4 US Department of Energy/EIA (2016) *International Energy Outlook 2016*. DOE/EIA-0484.
5 Straatman, P.J. and van Sark, W.G. (2008) A new hybrid ocean thermal energy conversion–Offshore solar pond (OTEC–OSP) design: A cost optimization approach. *Solar Energy*, 82 (6), 520–527.
6 Rehman, S., El-Amin, I.M., Ahmad, F., Shaahid, S.M., Al-Shehri, A.M., Bakhashwain, J.M. and Shash, A. (2007)

Feasibility study of hybrid retrofits to an isolated off-grid diesel power plant. *Renewable and Sustainable Energy Reviews*, 11 (4), 635–653.

7 Ball, M., Wietschel, M. and Rentz, O. (2007) Integration of a hydrogen economy into the German energy system: an optimizing modelling approach. *International Journal of Hydrogen Energy*, 32 (10), 1355–1368.

8 Shaahid, S.M. and Elhadidy, M.A. (2007) Technical and economic assessment of grid-independent hybrid photovoltaic–diesel–battery power systems for commercial loads in desert environments. *Renewable and Sustainable Energy Reviews*, 11 (8), 1794–1810.

9 Yilmaz, P., Hocaoglu, M.H. and Konukman, A.E. (2008) A pre-feasibility case study on integrated resource planning including renewables. *Energy Policy*, 36 (3), 1223–1232.

10 Greenpeace (2015) *Energy Revolution 2050. World Energy Scenario Report* (5th edn). Available at http://www.greenpeace.org/international/en/campaigns/climate-change/energyrevolution/

11 Cobb, L. (2008) *The Causes of Global Warming: A Graphical Approach.* http://tqe.quaker.org/2007/TQE158-EN-GlobalWarming.html

12 Ramanatharn, V., Ciceronhe, J., Singh, B. and Kiehl, T. (1985) Trace gas trends and their potential role in climate change. *Journal of Geophysical Research*, 90, 5547–5566.

13 Ipsos (2011) http://www.wpp.com/AnnualReports/2011/pdfs/wpp-ar11-full-report.pdf, p. 3.

14 Ellabban, O., Abu-Rub, H. and Blaabjerg, F. (2014) Renewable energy resources: Current status, future prospects and their enabling technology. *Renewable and Sustainable Energy Reviews*, 39, 748–764.

15 REN21 (2014) *Renewables 2014: Global Status Report*, pp. 13, 17, 21, 25. Available at http://www.ren21.net/Portals/0/documents/Resources/GSR/2014/GSR2014_full%20report_low%20res.pdf

16 O'Hayre, R., Cha, S.W., Colella, W. and Prinz, F.B. (2006) *Fuel Cell Fundamentals.* John Wiley & Sons, Inc., New Jersey.

17 US Department of Energy (2004) *Fuel Cell Handbook* (7th edn). DOE/NETL, Morgantown.

18 Srinivasan, S. (2006) *Fuel Cells: From Fundamentals to Applications*. Springer Science & Business Media.
19 DeBethune, A.J. (1955) Gibbs potentials as work functions. *Journal of the Electrochemical Society*, 50, 129–130.
20 Turns, S.R. (2006) *Thermodynamics*. Cambridge University Press, New York.
21 Larminie, J., Dicks, A. and McDonald, M.S. (2003) *Fuel Cell Systems Explained*. John Wiley, New York.
22 Williams, M.C. (2007) Solid oxide fuel cells: fundamentals to systems. *Fuel Cells*, 7 (1), 78–85.
23 Rokni, M. (1993) *Introduction of a fuel cell into a combined cycle: a competitive choice for future cogeneration.* In the 7th Congress & Exposition on Gas Turbines in Cogeneration and Utility Industrial and Independent Power Generation, Bournemouth, England. 09/21-23/93 1993, pp. 255–261.
24 Park, S.K., Oh, K.S. and Kim, T.S. (2007) Analysis of the design of a pressurized SOFC hybrid system using a fixed gas turbine design. *Journal of Power Sources*, 170 (1), 130–139.
25 Lundberg, W.L., Israelson, G.A., Moritz, R.R., Veyo, S.E., Holmes, R.A., Zafred, P.R., King, J.E. and Kothmann, R.E. (2000) *Pressurized Solid Oxide Fuel Cell/Gas Turbine Power System.* Federal Energy Technology Center, Morgantown, WV; Federal Energy Technology Center, Pittsburgh, PA.
26 Dias, J.A. and Assaf, J.M. (2008) Autothermal reforming of methane over Ni/γ-Al$_2$O$_3$ promoted with Pd: The effect of the Pd source in activity, temperature profile of reactor and in ignition. *Applied Catalysis A: General*, 334 (1), 243–250.
27 Pettersson, L.J. and Westerholm, R. (2001) State of the art of multi-fuel reformers for fuel cell vehicles: problem identification and research needs. *International Journal of Hydrogen Energy*, 26 (3), 243–264.
28 Bove, R. and Ubertini, S. (eds) (2008) *Modeling solid oxide fuel cells: methods, procedures and techniques.* Springer Science & Business Media.
29 Tanaka, K., Wen, C. and Yamada, K. (2000) Design and evaluation of combined cycle system with solid oxide fuel cell and gas turbine. *Fuel*, 79 (12), 1493–1507.
30 Holland, B.J., Zhu, J.G. and Jamet, L. (2007) *Fuel Cell Technology and Application.* University of Technology, Sydney, Australia.

31 Blomen, L.J. and Mugerwa, M.N. (eds) (2013) *Fuel cell systems.* Springer Science & Business Media.

32 Damo, U.M. (2016) *SOFC hybrid systems equipped with re-compression technology: transient analysis based on an emulator test rig.* PhD thesis, University of Manchester.

33 Hassmann, K. (2001) SOFC power plants, the Siemens-Westinghouse approach. *Fuel Cells*, 1 (1), 78–84.

34 Ciesar, J.A. (2001) *Hybrid systems development by the Siemens Westinghouse Power Corporation.* In US Department of Energy, Natural Gas/Renewable Energy Hybrids Workshop, Morgantown, WV, August, pp. 7–8.

35 Ferrari, M.L. (2006) *Transient analysis of solid oxide fuel cell hybrid plants and control system development.* PhD thesis, TPG-DiMSET, University of Genoa.

36 Stiller, C. (2006) *Design, operation and control modelling of SOFC/GT hybrid systems* PhD thesis, Norwegian University of Science and Technology Faculty of Engineering Science and Technology Department of Energy and Process Engineering.

2

SOFC Technology

Since the most interesting fuel cell technology for hybrid systems is based on either molten carbonate (MCFC) or solid oxide fuel cell (SOFC) stacks (significant devices for efficiency increase aspects), this chapter is mainly devoted to these components. Even if the discussion has to start from the basic aspects (including materials and component geometries), special attention is devoted to SOFC operations because they comprise the main component in the hybrid systems discussed in this book.

2.1 Basic Aspects of Solid Oxide Fuel Cells

Solid oxide fuel cells are based on a solid (ceramic) electrolyte able to operate in the 600–1000°C range. These are one of the most efficient devices [1] to convert fuel chemical energy

Hybrid Systems Based on Solid Oxide Fuel Cells: Modelling and Design, First Edition.
Mario L. Ferrari, Usman M. Damo, Ali Turan, and David Sánchez.
© 2017 John Wiley & Sons Ltd. Published 2017 by John Wiley & Sons Ltd.

directly to electrical energy with efficiency values in the 50–55% range. Since these devices produce power in direct current (DC) form, power electronic components (inverters) are necessary to connect SOFCs (as with the other fuel cell types) to standard electrical grids based on 50/60 Hz AC [1]. While up to 2005, SOFC technology was developed for very high temperature and efficiency conditions (900–1000°C), now special interest is also devoted to relatively low temperature SOFCs (600–900°C) to decrease costs related to high-temperature operations. While high temperatures allow for operation without noble-metal based catalysts (removing problems with CO poisoning), these operational conditions are extremely critical for thermal stresses on materials, including external vessels and connection pipes. However, high-temperature conditions allow for operation with a wide range of fuels. Although a significant hydrogen content is essential to operate an SOFC (the hydrogen reaction then has good kinetic performance), various hydrocarbons can be used as fuels [2]. This aspect allows electrochemical reactions directly on CO and CH_4, which are usually present in reformed gas flows. This is an important positive aspect as it removes the need for complex fuel cleaning devices, which are required in low-temperature fuel cell systems [3, 4]. Other positive aspects related to high-temperature conditions are linked with the following points: (i) operations with internal reforming, (ii) exhaust gases available as a heat source in a bottoming cycle; and (iii) important cogenerative applications [5] including the possibility of superheated steam generation for industrial processes.

There have been numerous studies concerning SOFC applications in different vehicle power systems (especially for large vehicles), but the main application for these fuel cells is stationary power generation (including applications in distributed generation grids). The suitability of SOFCs to stationary power generation applications is mainly due to the fairly long times required for the startup and shutdown phases related to operating at high temperature conditions without causing excessive thermal stresses to the stack. Since it is necessary to avoid high temperature gradients (usually it is necessary not to exceed 3 K/min [6]), startup/shutdown operations are not consistent with vehicle power systems. Moreover, the very high performance that can be obtained by matching SOFCs with microturbines (hybrid systems) shows important future perspectives to develop power generators able to obtain the highest electrical efficiency values (>65% on a low heating value (LHV) basis [1–6]).

2.2 SOFC Types

Even if past research activities on SOFCs were largely concerned with devices operating at 900–1000°C, interest is now also focused on SOFCs operating at lower temperature conditions due to the reduced cost of operation. For this reason, currently it is possible to classify this technology into the following categories: high-temperature and intermediate/low-temperature SOFCs.

2.2.1 High-temperature SOFCs

These are the conventional SOFCs based on yttria-stabilized zirconia (YSZ) electrolyte. Usually, the anode is a Ni–YSZ cermet and the cathode material is strontium-doped lanthanum (LSM) [7]. To achieve very high efficiency performance (or a high power density value), SOFCs falling into this category have to operate in the 800–1000°C temperature range. This high operating temperature causes very high costs for materials (interconnecting devices, external vessel, and pipes) and reliability concerns. These factors prevent widespread SOFC commercialization. Despite the high performance levels demonstrated by several high-temperature SOFC prototypes in the literature, currently special attention is devoted to solutions (including operational temperature decrease) necessary to obtain a significant cost decrease.

2.2.2 Intermediate/Low-temperature SOFCs

To decrease costs, low-temperature SOFCs were developed to operate in the 600–800°C range. This approach allows for a wide material choice for interconnecting devices, external vessels and pipes [8].

Moreover, a significant temperature decrease to values lower than 650°C would generate the following advantages linked with the cost decrease [8]:

- Low-cost metallic materials (e.g. ferritic stainless-steels) for the interconnections and construction materials. This produces a very cheap and robust balance-of-plant and stack contamination aspects linked with chromium-based materials.
- Fast startup and shutdown phases.
- Simplification of the design and materials aspects for the balance-of-plant.
- Significant corrosion rate decrease (reliability increase).

To decrease SOFC operating temperatures, researchers have considered the following approaches:

- Electrolyte thickness decrease for the YSZ conventional material [9];
- Alternative electrolyte materials with higher conductivity in low-temperature conditions (e.g. doped ceria or lanthanum gallate [10]);
- Electrode voltage drop decrease [11].

Many techniques have been studied and developed for producing thin-film electrolytes with the objective of reducing electrolyte resistance in low-temperature conditions [12]. Important advances were obtained on these types of reactors based on a thin-film doped ceria electrolyte (mainly gadoline-doped ceria in combination with several cathode types) [13]. Also, important efforts have been carried out to improve the cathode performance and materials to achieve the desired power density objective: $400 \, mW/cm^2$ peak power density at $500°C$ [13]. Moreover, several related studies were carried out (or are under development) to improve low-temperature SOFC performance [8]

2.3 Materials for SOFCs

Various property considerations have to be taken into account prior to selecting the appropriate materials when designing and developing the electrolyte, cathode, and anode of an SOFC. These considerations primarily refer to the specific electrochemical behaviour, including thermal expansion, stability, and oxygen ionic and electronic transport properties. Judiciously matching these required component attributes is the hardest step in designing solid oxide fuel cells [14].

These properties are especially important at the triple phase boundary (TPB) where the electrolyte, electrode and gas interact. At the cathode–electrolyte interface, oxygen disassociates and is reduced to oxygen ions that are conducted across the electrolyte, while at the electrolyte–anode interface the fuel is electrochemically oxidized with the oxygen ions. Composite electrodes incorporate materials similar to the electrolyte layer which allow for better mechanical property matching (such as thermal expansion coefficients) and improved reaction kinetics [14].

As mentioned in the previous section, yttria-stabilized zirconia (with 3–8% or 10% of yttrium content), referred to as YSZ, is the most common electrolyte for SOFCs [15]. However, for efficiency enhancement and operations at lower temperatures, the following alternative electrolyte types were proposed: scandium-doped zirconia (SDZ) (more conductive than YSZ), gadolinium-doped ceria (even more conductive, but partially reduced at temperatures above 600°C), lanthanum gallate with strontium doping (the A-site of perovskite) and magnesium (the B-site) (for temperatures as low as 600°C). However, these alternative electrolytes are still under development, with attendant thermal expansion problems [16].

For the anodic electrode, the most commonly used material is a cermet of nickel with YSZ [16]. Nickel is mainly used as the anode material due to its high electrical conductivity and catalytic activity in the fuel-oxidation reaction. It is normally blended with yttria-stabilized zirconia (YSZ) to form a cermet. A particular reason for this choice involves matching the thermal expansion properties of the chosen electrolyte, as Ni is a poor ionic conductor. Adding YSZ also allows the reaction zone to extend further into the anode from the TPB, due to its ionic conduction properties [14].

Currently, several research activities are being developed to mitigate the following problems: sulphur sensitivity, oxidation reduction intolerance, stability problems during thermal cycling, and poor activity for hydrocarbon oxidation. Among the different efforts to improve the anode, it is prudent to highlight that new materials proposed by the Pacific Northwest National Lab (PNNL) demonstrated an interesting sulfur tolerance (up to 100 ppm) and good oxidation/reduction stability.

For the cathode, the most commonly used materials are lanthanum-based perovskites due to the following desirable attributes [17]: chemical stability, adequate conductivity, chemical activity, interaction with the ceramic material, and thermal expansion coefficients similar to those of the electrolyte. For high-temperature SOFCs, strontium-doped $LaMnO_3$ (LSM) is also used, while for intermediate temperature devices a composite layer of YSZ and LSM is often used. Certainly, strontium-doped lanthanum manganite (LSM) is the material of choice [14]. Like the Ni anode, LSM is a poor ionic conductor and is usually paired with a good ionic conductor to expedite the cathode reactions [14]. YSZ can again be used, but other materials such as gadolinium-doped ceria (GDC) are possible as well.

GDC is a better ionic conductor than YSZ and can be employed in both the cathode and anode. In the anode, a GDC interlayer can be employed between the dense electrolyte and the typical Ni–YSZ cermet to improve the anode reaction kinetics and ionic conductivity [14]. In the cathode, GDC can also be used instead of YSZ to increase the ionic conductivity, with the added benefit of being unreactive with LSM [14].

Materials for interconnecting components are especially critical for high-temperature SOFCs due to high cost considerations [18]. Specifically, for operations in the 900–1000°C range, conductive ceramic materials are used (the most used materials are the doped lanthanum and yttrium chromites including Mg, Sr, Ca, Ca/Co as dopants). However, these costs can be strongly reduced if temperatures are decreased to allow application of ferritic steels (materials easy to process with low-cost techniques, in comparison with ceramics) [19].

2.4 Different SOFC Geometries

Two main geometries are in use for SOFCs: tubular and planar layouts. In comparison with the geometries considered for the other fuel cell types, the tubular layout is unique to SOFCs.

2.4.1 Tubular SOFCs

The tubular geometry is based on a cell design similar to a laboratory test tube. The anode–electrolyte–cathode layers are shown in detail in Figure 2.1. Internal to the SOFC tube, an air heating tube (quartz or Al_2O_3) is included [4, 20–22]; thus, while air flows in the internal ducts (the heating tube is upstream of the cathodic side), fuel is flowing externally (anodic side). This is then categorized as a co-flow fuel cell because both flows are moving from the bottom to the top in Figure 2.1. Downstream of the cell tube, the flows through both sides are mixed in the off-gas burner zone before leaving the system as exhaust gas. For the current YSZ electrolyte, it is necessary to operate at high temperature (900–1000°C range) to provide enough oxygen conductivity. Therefore, expensive (high temperature) alloys have to be used for the fuel cell housing, resulting in a significant cost increase. However, since significant cost decrease can be obtained with temperatures in the 600–800°C range, special attention is now devoted to materials able to operate with good conductivity in this low temperature condition [23].

Since the fuel cell is based on a complex geometry, standard manufacturing processes are not available. Therefore the tube (the cathode of the cell) fabrication is carried out with extrusion and sintering methods and the electrolyte is applied to the tube using electrochemical vapour deposition. The anode is generated with a slurry deposition on the electrolyte and

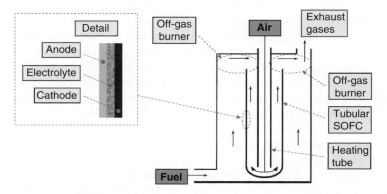

Figure 2.1 Schema for a tubular SOFC.

the interconnecting material is formed with a plasma spray approach [23].

Since several manufacturers have developed tubular SOFCs based on different approaches, the following discussion provides the main details and differences related to these fuel cells.

2.4.1.1 Electrical Conduction Around the Tube

This is the most significant approach developed as it was chosen by Siemens-Westinghouse to provide a reference for the development of tubular SOFC technology (including the first hybrid system). In this configuration, each tube has an external interconnector device along the entire element to operate a complete electrical connection between the tubes (either in series or in parallel) [23].

2.4.1.2 Electrical Conduction Along the Tube

In this case, electrical current is conducted axially (along the tube) and interconnectors are applied just on the top of the tubes [23].

The primary manufacturer responsible for SOFCs with this electrical connection (such that the conduction is along the tube) is Acumentrics, who is specialized in producing 2 kW SOFC stacks.

2.4.1.3 Segmented-in-series Tubular SOFCs

These types of tubular SOFCs are promising for the reduction of manufacturing costs as the complex geometry of the tube is substituted with a segmented cell area generated along the support (the tube). The cells are connected in series from the electrical point of view using the following approach: the anode of a cell is connected to the cathode of the following one [24]. Figure 2.2 shows the layout of this configuration. The approach displayed is a geometry modification developed to reduce costs of the tubular SOFC technology, and is quite different in layout from the schema in Figure 2.1. In comparison with the tubular cell types discussed in the previous subsection, this segmentation results in a higher voltage per tube, thus allowing the cell to operate with a lower current density and thus obtain a significant service lifespan increase. The main manufacturer involved in developing this SOFC type is Mitsubishi Heavy Industries (MHI) for operations at both atmospheric and pressurized conditions.

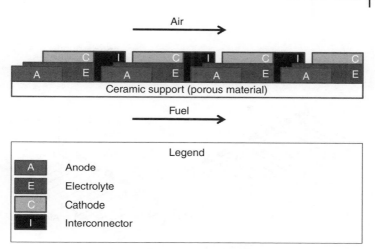

Figure 2.2 Schema for a segmented-in-series tubular SOFC.

Another SOFC design that can be classified a segmented-in-series tubular SOFC is the IP-SOFC design developed and produced by Rolls-Royce Fuel Cell Systems. It is based on a flattened tube configuration with electrical connections carried out as in the segmented-in-series cells [25, 26].

2.4.2 Planar SOFCs

Planar SOFCs were developed to pursue the cost decrease objective as they are based on a simpler design to manufacture [9, 10] in comparison with the tubular cell. In this case, the cell components are flat plates connected in series from the electrical point of view. Figure 2.3 shows the basic unit for these kinds of cells.

Planar SOFCs are usually classified into the following categories: self-supporting and external-supporting layouts. In the self-supporting category, usually the thickest layer is designed to provide the structural strength. So, these kinds of cells are usually referred to as anode-supported, electrolyte-supported or cathode-supported planar SOFCs. The electrolyte-supported cells are usually based on a 100 μm YSZ electrolyte (for operations in the 900–1000°C range), while the anode-supported SOFCs are usually equipped with 5–20 μm electrolyte (for operations at temperature values lower than 800°C) [23].

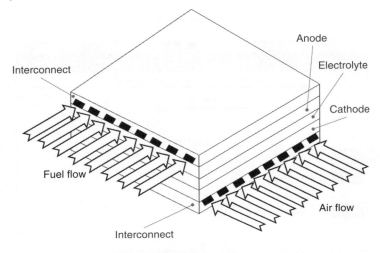

Figure 2.3 Schema for a planar SOFC.

These kinds of fuel cells are manufactured based on the following main flow path layouts: cross-flow, co-flow or counter-flow. However, since the flow layout is linked with the temperature distribution, to increase the thermal uniformity (reducing stress on materials), alternative flow patterns are considered: Z-flow, serpentine, spiral and radial flow layouts. Moreover, for efficient SOFC connections with microturbines, it is also necessary to select a flow pattern able to avoid too much pressure loss. So, a detailed fluid dynamic analysis is essential to define the optimal planar SOFC geometry [7].

Since each cell is connected with manifolds to distribute the flows to the internal ducts, SOFC classification can also be based on the specific configuration of this component: external or integral manifolds [7]. While the first solution is based on components completely separated from the cell, the latter is related to manifold geometries integrated with the cell parts. In both cases, manifolds are designed to prevent gas leakage or crossover, to avoid electrical short-circuits, to minimize pressure losses and to provide uniformity in the flow distribution.

2.5   SOFC Stacks

Since voltage and power related to a single SOFC are very low, usually single cells are stacked to generate a device able to operate at about 50 V (upstream of the electrical power conditioning systems), producing a power value higher than 1 kW (100 kW or 250 kW are typical SOFC sizes).

In the case of tubular SOFCs the cells are stacked as shown in Figure 2.4 and connected in series and in parallel (batches of cells connected in series linked together in parallel) to obtain a significant voltage close to 50 V (or other similar values according to the power electronic requirements) [23].

The stacks composed of planar cells are manufactured simply by putting one cell element (Figure 2.3) over the next [23]. Usually, the stack is equipped with external tie-rods to compress the elements with the objective of minimizing contact resistance and

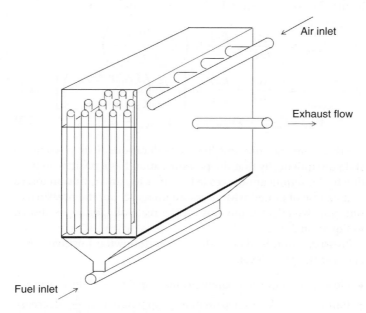

Figure 2.4 Stack based on tubular SOFCs.

flow leakages. However, this is a very critical manufacturing step to be carried out with precise torque metering because too much mechanical stress can easily damage the entire stack.

2.6 Effect of Pressurization for SOFCs

From Equation 2.1, considering a simple hydrogen–oxygen electrochemical reaction, it is possible to highlight the positive influence of the pressure increase on the Nernst's voltage (the ideal voltage) for the SOFCs. Moreover, since the logarithm argument is proportional to the square root of the pressure, the most significant voltage increase is related to an operational pressure change from atmospheric conditions to 4–7 bar. Even if the voltage value continues to increase with further pressure increase, usually the benefit obtained is not so significant to justify the cost related to high-pressure conditions.

$$
\begin{aligned}
E_{ideal} &= -\frac{\Delta G^0}{n\cdot F} + \frac{R\cdot T}{n\cdot F}\ln\left(\frac{(p_{O_2})^{1/2}\cdot p_{H_2}}{p_{H_2O}}\right) \\
&= -\frac{\Delta G^0}{n\cdot F} + \frac{R\cdot T}{n\cdot F}\ln\left[p^{1/2}\cdot\left(\frac{(\chi_{O_2})^{1/2}\cdot \chi_{H_2}}{\chi_{H_2O}}\right)\right] \quad (2.1)
\end{aligned}
$$

$$
V_{real} = E_{ideal} - \Delta V_{activation} - \Delta V_{ohmic} - \Delta V_{diffusion} \quad (2.2)
$$

Also, from the inherent loss considerations, it is possible to state unequivocally that the pressurization has a positive benefit due to the significant decrease in activation and diffusion losses that can be obtained with pressure increase. So, the pressurization positive effect is obvious on the real voltage too, as shown in Equation 2.2.

Pressurization is also related to the following positive influences at the plant level:

- Flow losses (approximately proportional to $\frac{1}{p^{1/2}}$) decrease;
- Pumping work (approximately proportional to $\frac{1}{p^2}$) decrease due to a power density increase;
- Heat exchanger area (approximately proportional to $\frac{1}{p^{0.5}} \div \frac{1}{p^{0.8}}$) and cost decrease.

Moreover, in the case of pressurized hybrid systems based on a recuperated turbine coupled with the SOFC, the pressurization is linked with lower temperature conditions on the recuperator. This allows for the use of stainless steel instead of high-cost materials, and thus a significant positive effect on the total system costs.

2.7 Fuel Processing for SOFCs

SOFCs are able to accept different fuel types including CH_4 and CO; however, due to kinetic aspects, it is not possible to efficiently operate these electrochemical reactors directly with standard commercial fuels [23]. For this reason, for natural gas (as for other fuel types) it is necessary to consider fuel processing techniques to produce hydrogen-rich gases. Depending on the fuel type (especially on the fuel phase), different approaches can be considered for the selection and installation of appropriate reactors. Despite the fact that fuel processing reactions can be carried out upstream of the plant fuel intake duct, integrated configurations are worth consideration as they have the added benefit of being able to exploit the heat generated by SOFC reactions and/or the steam content available at the exhaust ducts. To further elaborate, there is a reduction in the cost of operation associated with the use of a proportion of less refined fuel types, and it is beneficial to make use of the heat content produced by SOFC reactors in this process. To achieve this, some plant layouts are based on a small fuel processing reactor (usually a pre-reformer for natural gas fed systems), while the majority of fuel processing reactions are carried out internally in the SOFC. This approach, typically named internal reforming, has two different basic configurations: (i) indirect internal reforming (IIR) in the case the reactor is located in thermal contact with the fuel cell (but separated from it); (ii) direct internal reforming (DIR) if the fuel processing reactions are carried out directly inside the anodic ducts of the SOFC stack [23].

Since several resources [1, 7, 23] provide details of the relevant fuel processing techniques, the following subsections will only give a brief summary including the aspects necessary for hybrid systems. Due to the differences in reactor types, complexity and gas treatment, this discussion is divided into two subsections: the processing of gas and liquid fuels, and the processing of solid fuels (especially coal).

2.7.1 Processing for Gas and Liquid Fuels

Equation 2.3 shows the general chemical reaction for processing techniques related to both gas and liquid fuels (z is the oxygen-to-fuel molar ratio).

$$C_nH_mO_p + z\left(O_2 + \frac{79}{21}N_2\right) + (n - 2{\cdot}z - p)H_2O \rightarrow nCO$$
$$+ \left(n - 2{\cdot}z - p + \frac{m}{2}\right)H_2 + \frac{79}{21}zN_2 \qquad (2.3)$$

Considering the value of z, it is possible to define the following fuel processing techniques [23]: (i) steam reforming if $z = 0$; (ii) partial oxidation if $z = 1$; (iii) autothermal reforming if z has an intermediate value between 0 and 1.

2.7.1.1 Steam Reforming

This is the most popular and developed technique to convert light hydrocarbon fuels into a hydrogen-rich gas mixture. To operate the conversion with this process the fuel has to be heated, vaporized and injected (with superheated steam) into a reactor equipped with a catalyst (usually nickel, but cobalt and noble metals can be used at additional cost). Since steam reforming is an endothermic reaction, it is favoured by high temperatures. So, a significant heat input is necessary to the flow by pre-heating, utilizing an adjacent furnace or the heat released by the fuel cell. Optimal values of temperature, steam-to-carbon ratio and catalyst type are dependent on the fuel to be processed [23]. As an example of steam reforming, the reaction related to methane is given as (Equation 2.4):

$$CH_4 + H_2O \rightarrow CO + 3H_2 \qquad (2.4)$$

Steam reforming is a well consolidated technique for gas fuels due to the easy mixing between the flows and the achievable high efficiency of conversion performance [23]. For this reason, this process is used extensively in natural gas fed hybrid systems.

2.7.1.2 Partial Oxidation

This process is based on a substoichiometric fuel/oxygen mixture to obtain partial combustion. This approach is exothermic, producing a high temperature condition for the products, and quenching could be necessary to have the fuel at the right temperature. There are two main categories of technology using partial oxidation for fuel processing: (i) non-catalytic partial oxidation (POX) and (ii) catalytic partial oxidation (CPOX). The application of a catalyst increases the reactor performance (e.g. allowing reactor size decrease in comparison with non-catalytic units) and decreases the operating temperature. Typical temperature values of partial oxidation reactors range from 1400°C for POX down to 870°C for CPOX technology [23].

As an example of partial oxidation, this is the reaction related to methane (neglecting nitrogen as diluent) (Equation 2.5):

$$CH_4 + \tfrac{1}{2}O_2 \rightarrow CO + 2H_2 \tag{2.5}$$

As can be highlighted by comparing equations 2.4 and 2.5, for gas fuels the partial oxidation technique produces a smaller amount of hydrogen in comparison with the steam reforming process. However, due to flow mixing aspects and efficiency reasons, partial oxidation is considered a better fuel processing solution for liquid fuels, even if (at the moment) the technology is not as well consolidated as steam reforming for natural gas [23].

2.7.1.3 Autothermal Reforming

This is an advanced solution to conjugate the positive aspects of steam reforming (SR) and partial oxidation (POX) techniques [23]. In autothermal reforming, SR and POX reactions occur in the same reactor without a physical separation between the zones, to utilize the heat produced by the exothermic POX reaction in the endothermic SR reaction. This results in a slightly exothermic process. In comparison with a simple SR approach, the utilization of an autothermal reforming (ATR) technique allows for operation with a more compact reactor. ATR is able to perform faster at startup, react faster to load change phases, and produce a more hydrogen-rich fuel [27]. However, in spite of several research activities on this technology, at the moment ATR is considered less mature or reliable than SR and POX.

2.7.2 Processing for Solid Fuels

A hydrogen-rich gas can be obtained from solid fuels by utilizing a gasification process. This technology can be applied to different solid fuels (such as coal, biomass, etc.). However, this subsection devotes special attention to coal processing as it is extensively used and an important interest in numerous technologies [28].

To produce a syngas ('synthesis gas'), coal (after a pre-treatment) is mixed with steam and oxygen (it can be pure oxygen obtained using an air separation unit or simply an air flow). The main reaction (Equation 2.6), named gasification, is coupled with the reactions of the partial and total combustion of coal and hydrogen.

$$C + H_2O \rightarrow H_2 + CO \tag{2.6}$$

Three main choices of reactor technologies are available for coal gasification: moving bed, fluidized bed and entrained bed [23]. Moving bed reactors are based on a mature technology able to operate at moderate temperature conditions (425–650°C) and at pressure values in the 1–20 bar range. The produced syngas contains a large amount of light hydrocarbons (CH_4, C_2H_6), but also H_2S, COS, NH_3, HCN, naphtha, tar, fuel oil and phenols. The reactors based on fluidized bed technology are less widespread from a commercialization point of view. They operate in the 925–1040°C temperature range and for pressure values from 1 bar to 100 bar in advanced applications. The produced syngas with fluidized bed technology contains significantly lower amounts of H_2S, COS, NH_3, HCN, naphtha, tar, fuel oil and phenols in comparison with that produced by moving bed reactors. Finally, entrained bed reactors are able to operate at very high temperature conditions (>1260°C) and in the 1–300 bar pressure range. They typically produce a syngas almost entirely composed of H_2, CO and CO_2, and thus provide a significant advantage in terms of the reduced need for gas cleaning components and resulting cost decrease. Moreover, these reactors are small, compact and able to operate on all coal types. Despite these advantages, there are currently only a limited number of entrained bed applications as the technology is relatively immature with important unresolved issues, mainly related to the high-temperature operating conditions.

2.7.2.1 Syngas Treatment

It is often essential to install further reactors [23] to increase the amount of hydrogen in the syngas and/or to remove composites or elements able to produce pollutants or damage the SOFC components.

An important process to achieve this aim is water-gas shifting (Equation 2.7), which is usually necessary specifically to increase the hydrogen content and reduce the levels of CO. This reaction can be directly performed inside the main fuel processing reactor, but in some configurations devoted devices are installed to improve the conversion performance.

$$CO + H_2O \rightarrow H_2 + CO_2 \qquad (2.7)$$

Additionally, depending on the fuel type, it could be necessary to install desulfurizer reactors and/or particulate removal devices. Sulfur is usually removed in the form of H_2S in low-temperature processes using chemical or physical absorption, and/or utilizing zinc-ferrite and zinc-titanate for high-temperature devices. Furthermore, the particulates could be removed at low-temperature conditions using cyclones and/or electrostatic precipitators, or at high-temperature conditions using ceramic devices (not currently ready for the market due to low reliability and performance).

Finally, since the produced syngas has also to be managed from the temperature point of view, the gas treatment line could also be equipped with heat exchangers and/or with water quenching devices.

2.8 SOFC Applications in Hybrid Systems

Since SOFCs operate at high temperature conditions, they produce an exhaust flow with a significant energy content. Therefore, a possible coupling with standard energy systems is considered a promising solution for maximizing plant electrical efficiency (and hence a positive effect on CO_2 release and other pollutant emission reductions). A number of system integration solutions have been designed involving, for example, steam power plants [29] and combined cycles [29]. However, the most promising approach for a hybrid system prototype couples an SOFC stack with a micro gas turbine. This is partly due to the fact that large SOFC systems to be coupled with steam or combined plants are not, at the moment, developed adequately, because the small-size plant design and development issues have not been resolved satisfactorily. Furthermore, the development of real systems based on the coupling of SOFCs with steam or combined plants with small-scale SOFC systems has significant hurdles to overcome with respect to efficiency. Future interest and development of large hybrid systems including steam turbines could be linked with the improvement of intermediate/low-temperature SOFCs.

Hybrid systems based on the coupling between SOFCs and microturbines can be classified into two categories: atmospheric systems and pressurized systems.

2.8.1 Atmospheric SOFC Hybrid Systems

In atmospheric hybrid systems, the SOFC is operated at atmospheric conditions as shown in Figure 2.5 [7, 29]. The cathodic inlet of the cell receives hot air from the expander outlet duct. On the anodic side, the system can operate with pure hydrogen (as in Figure 2.5) or with natural gas embodying

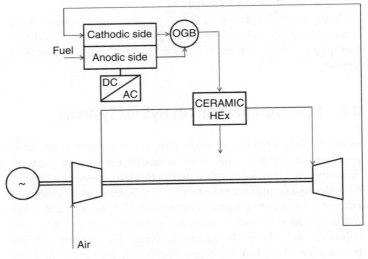

Figure 2.5 General scheme of an atmospheric SOFC hybrid system.

reforming reactions. In this case, anodic recirculation could be an interesting solution, even though it is not given in Figure 2.5 for the sake of simplicity. The connection between the stack and the turbine is carried out with a high-temperature heat exchanger [29]. To operate at high efficiency conditions, it is necessary to reach high temperature values (in this component), which are usually not sustainable by standard heat exchangers. It has been proposed to use ceramic devices to overcome this problem. This solution results in a significant cost increase, and the high cost aspects related to this component are not compensated for by efficiency performance increase, because SOFCs at atmospheric conditions have lower voltage values in comparison with pressurized solutions. Furthermore, since currently ceramic heat exchangers have not reached sufficient reliability performance levels, these types of hybrid systems have significant obstacles to overcome, and development interest largely prioritizes pressurized systems.

2.8.2 Pressurized SOFC Hybrid Systems

In pressurized hybrid systems the SOFC stack is located between the compressor outlet and the expander inlet, as shown in Figure 2.6. The cathodic side of the cell receives compressed air from the compressor. At this point, different plant layouts are possible: (i) systems equipped with a recuperator [7, 22] as in Figure 2.6; (ii) systems equipped with a cathodic recirculation as the only device for air pre-heating [25, 26]; (iii) systems equipped with both a recuperator and cathodic recirculation; and (iv) plant layouts based on a pre-heating burner [7, 29]. While the first solution can only be used with tubular SOFCs (e.g. the Siemens-Westinghouse plant [7, 22]) because the air pre-heat has to be completed with the quartz internal tube, the other solutions can also be applied to flattened-tube (e.g. the Rolls-Royce Fuel Cell Systems plant [25, 26]) or planar cell layouts. Further details related to these configurations are reported in [21]. On the anodic side, the system can operate with pure hydrogen (as in Figure 2.6) or with natural gas toward reforming reactions. In this case, anodic recirculation could be an interesting solution. In comparison with the atmospheric layout, no high temperature and expensive heat exchangers are used and a significant performance increase due to pressurization is obtained.

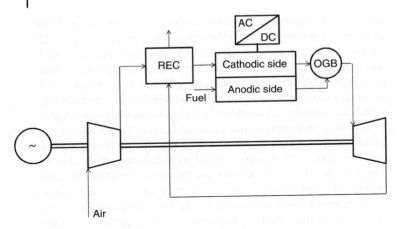

Figure 2.6 General scheme of a pressurized SOFC hybrid system.

Moreover, the cost increment due to pressurization is not too high because, as discussed previously, a pressure value in the 4–7 bar range is adequate to obtain a significant efficiency increase.

2.9 Aspects Related to SOFC Reliability, Degradation and Costs

Since the applicability of SOFC-based plants to wide-scale commercialization is strongly linked with their reliability, degradation and costs, it is important to consider these aspects carefully. Specifically, in comparison with traditional power plants, a higher efficiency performance generates a significant marginal cost reduction (due to fuel utilization decrease) that can compensate for a higher component cost or lower reliability (due to faster degradation of SOFC components). However, to penetrate the market, the total cost (taking into account the additional maintenance due to SOFC degradation) has to be lower than the typical values in traditional power-generating plants. Even if some benefit for SOFC technology development can be obtained from incentives related to low emissions, the market issues still have to be overcome. Simply stated, incentives can move the market to SOFC-based systems by compensating for some additional costs, but they will be not able to compensate a doubled (or higher) total cost.

A study carried out by the US Department of Energy [30] calculated the estimated capital costs related to a 270 kW fuel cell system based on a planar SOFC stack. The latest and lowest cost manufacturing technologies [30] were considered, including the balance of plant costs for a stationary distributed generation system. The Equation 2.8 was developed for the total capital costs (y expressed in \$/kW) on the basis of the annual production (x):

$$y = 12900 \cdot x^{-0.332} \tag{2.8}$$

Even if a large-scale fuel cell production can generate a significant cost decrease, these values remain higher than the related costs for standard technologies. For such reasons, other aspects including incentives, electricity and fuel prices, electrical efficiency and fuel cell degradation performance are very important considerations for market penetration scenarios related to these systems.

Due to the importance of SOFC degradation regarding system cost and market penetration, special attention is devoted here to this topic. This negative aspect is related to the electrochemical performance decay (voltage decrease) caused by several issues such as poisoning coming from fuel impurities, coarsening of the nickel, contaminants in the triple phase zone, microstructural change, delamination, chromium migration, and so on. An example of possible analyses on SOFC degradation (and the related impact on hybrid systems) was carried out by the National Energy Technology Laboratory (NETL, US Department of Energy) [31] on the basis of experimental results from several research groups [32]. The fuel cell considered by NETL was an anode-supported SOFC with an YSZ anode, a LSM and LSM/YSZ interlayer-based cathode and an electrolyte composed of YSZ material. The cathode side was fed by air, and syngas (H_2: 29.1%; CO: 28.6%; H_2O: 27.1%; CO_2: 12.0%; N_2: 3.2%) was considered for the anode side. The Equation 2.9 was obtained with an extrapolation using an algorithm from SigmaPlot™. Even if the experimental data were not detailed enough to produce a precise predictive equation, a set of curves was obtained considering an exponential trend in the degradation rate (named r_d in Equation 2.9 in percentage of voltage drop per hour) with respect to current density (i) in A/cm^2. FU is the fuel utilization factor, T is the cell temperature (in Kelvin) and t is time (in hours) [31].

$$r_d = \frac{0.59 \cdot FU + 0.74}{1 + e^{\frac{T-1087}{22.92}}} \cdot (e^{2.64 \cdot i} - 1) \cdot \frac{t}{1000} \qquad (2.9)$$

2.10 Conclusions

SOFC technology is of great interest in hybrid system development, especially for both high efficiency and low emission performance. This chapter started from the basic aspects relating to SOFC component materials, SOFC geometries, cell stacking and fuel processing, then attention was devoted to hybrid system development. A simple classification of SOFC-based hybrid systems was then presented to introduce the detailed discussions on these innovative plants included in the next chapters. Since the potential for wide-scale commercialization of SOFC-based plants is strongly linked to their reliability, degradation and cost aspects, the results obtained by the US Department of Energy's investigation into cost and degradation were reported and analyzed.

2.11 Questions

1 How can SOFCs be classified considering the operating temperature?

2 What is the function of the air heating tube in tubular SOFCs?

3 What are the main differences between tubular and planar SOFCs?

4 What is the main critical component in atmospheric hybrid systems?

5 What is the ideal voltage increase for a SOFC pressurized at 4 bar in comparison with 1 bar operations? Consider the following data:

- SOFC average temperature: 900°C;
- chemical concentrations: equal in both cases;
- inlet flows: pure H_2 and pure O_2.

6 What are the main fuel processing techniques for producing a hydrogen-rich syngas?

7 Why is SOFC degradation very important for market penetration related to hybrid systems?

8 What are the main causes of SOFC degradation?

References

1 Singhal, S.C. and Dokiya, M. (2003) *Solid Oxide Fuel Cells VIII*. The Electrochemical Society Proceedings, Pennington, NJ, PV2003-07.

2 Birgersson, K.E., Balaya, P., Chou, S.K. and Yan, J. (2012) Energy solutions for a sustainable world. *Applied Energy*, 90, 1–2.

3 Richards, G.A., McMillian, M.M., Gemmen, R.S., Rogers, W.A. and Cully, S.R. (2001) Issues for low-emission, fuel-flexible power systems. *Progress in Energy and Combustion Science*, 27, 141–169.

4 Veyo, S., Shockling, L.A., Dederer, J.T., Gillett, J.E. and Lundberg, W.L. (2002) Tubular solid oxide fuel cell/gas turbine hybrid cycle power systems: Status. *Journal of Engineering for Gas Turbines and Power*, 124, 845–849.

5 Lilley, P.D., Erdle, E. and Gross F. (1989) *Market Potential of SOFC*. Report No. EUR 12249 EN. CEC, Luxembourg.

6 Ferrari, M.L., Pascenti, M., Magistri, L. and Massardo, A.F. (2010) Hybrid system test rig: start-up and shutdown physical emulation. *Journal of Fuel Cell Science and Technology*, 7, 021005_1-7.

7 Singhal, S.C. and Kendall, K. (2003) *High Temperature Solid Oxide Fuel Cells: Fundamentals, Design and Applications.* Elsevier Advanced Technology, Oxford.

8 Liu, Q.L., Khor, K.A. and Chan, S.H. (2006) High-performance low-temperature solid oxide fuel cell with novel BSCF cathode. *Journal of Power Sources*, 161, 123–128.

9 De Souza, S., Visco, S.J. and De Jonghe, L.C. (1997) Thin-film solid oxide fuel cell with high performance at low-temperature. *Solid State Ionics*, 98, 57–61.

10 Fukui, T., Ohara, S., Murata, K., Yoshida, H., Miura, K. and Inagaki, T. (2002) Performance of intermediate temperature solid oxide fuel cells with La(Sr)Ga(Mg)O$_3$ electrolyte film. *Journal of Power Sources*, 106, 142–145.

11 Xia, C., Rauch, W., Chen, F. and Liu, M. (2006) Sm$_{0.5}$Sr$_{0.5}$CoO$_3$ cathodes for low-temperature SOFCs. *Solid State Ionics*, 149, 11–19.

12 Will, J., Mitterdorfer, A., Kleinlogel, C., Perednis, D. and Gauckler, L.J. (2000) Fabrication of thin electrolytes for second-generation solid oxide fuel cells. *Solid State Ionics*, 131, 79–96.

13 Steele, B.C.H. (2000) Appraisal of Ce$_{1-y}$Gd$_y$O$_{2-y/2}$ electrolytes for IT-SOFC operation at 500°C. *Solid State Ionics*, 129, 95–110.

14 Ross, T.B.A. (2015) Performance and stability of large planar solid oxide fuel cells using phosphine contaminated hydrogen fuel. MSc thesis, West Virginia University, Morgantown, WV.

15 Wang, S., Kobayashi, T., Dokiya, M. and Hashimoto, T. (2000) Electrical and ionic conductivity of Gd-doped ceria. *Journal of Electrochemical Society*, 147 (10), 3606.

16 Arachi, Y., Sakai, H., Yamamoto, O., Takeda, Y. and Imanishi, N. (1999) Electrical conductivity of the ZrO$_2$–Ln$_3$O$_3$ (Ln = lanthanides) system. *Solid State Ionics*, 121, 133–139.

17 Möbius, H.-H. (2003) *History in High Temperature Solid Oxide Fuel Cells: Fundamentals, Design and Applications.* Elsevier, Oxford, pp. 23–51.

18 Singhal, S.C. (1997) *Recent progress in tubular solid oxide fuel cell technology.* Proceedings of the Fifth International Symposium on Solid Oxide Fuel Cells (SOFC – V), The Electrochemical Society, Inc., Pennington, NJ.

19 Yang, G.Z., Xia, G., Singh, P. and Stevenson, J. (2004) *Advanced Metallic Interconnect Development*. In SECA 2004 Annual Meeting and Core Program Review. US DOE NETL, Boston.

20 Zhao, H., Dang, Z. and Xi, G. (2011) Investigation of coupling characteristics of SOFC/MG hybrid 418 system. *Journal of Engineering Thermophysics*, 32, 1647–1650.

21 Ferrari, M.L. and Massardo, A.F. (2013) Cathode–anode interaction in SOFC hybrid systems. *Applied Energy*, 105, 369–379.

22 Ferrari, M.L. (2011) Solid oxide fuel cell hybrid system: control strategy for stand-alone configurations. *Journal of Power Sources*, 196, 2682–2690.

23 US Department of Energy (2004) *Fuel Cell Handbook* (7th edn). DoE, Morgantown, WV.

24 Mitsubishi Heavy Industries, Ltd. (2007) Technical Review, Vol. 44, No. 1, March 2007.

25 Agnew, G.D., Bozzolo, M., Moritz, R.R. and Berenyi, S. (2005) *The design and integration of the Rolls-Royce Fuel Cell Systems 1MW SOFC*. ASME Paper GT2005-69122.

26 Trasino, F., Bozzolo, M., Magistri, L. and Massardo, A.F. (2011) Modeling and performance analysis of the Rolls-Royce Fuel Cell systems limited: 1 MW plant. *Journal of Engineering for Gas Turbines and Power*, 133, 021701_1–11.

27 Mancusi, E., Acampora, L., Marra, F.S. and Altimari, P. (2015) Hysteresis in autothermal methane reforming over Rh catalysts: Bifurcation analysis. *Chemical Engineering Journal*, 262, 1052–1064.

28 Seyitoglu, S.S., Dincer, I. and Kilicarslan, A. (2016) Assessment of an IGCC based trigeneration system for power, hydrogen and synthesis fuel production. *International Journal of Hydrogen Energy*, 41 8168–8175.

29 Hirschenhofer, J., Stauffer, D. and Engleman, R. (1994) *Fuel Cells: A Handbook (Revision 3)*. Gilbert/Commonwealth, Inc. for the US Department of Energy under Contract No. DE-ACO1-88FE61684.

30 Weimar, M., Chick, L., Gotthold, D. and Whyiatt, G. (2013) *Cost Study for Manufacturing of Solid Oxide Fuel Cell Power Systems*. Pacific Northwest National Laboratory, US Department of Energy.

31 Zaccaria, V., Tucker, D. and Traverso, A. (2016) A distributed real-time model of degradation in a solid oxide fuel cell, part I: Model characterization. *Journal of Power Sources*, 311, 175–181.

32 Hagen, A., Barfod, R. and Hendriksen, P.V. (2006) Degradation of anode supported SOFCs as a function of temperature and current load. *Journal of Electrochemistry Society*, 153, A1165–A1171.

3

Micro Gas Turbine Technology

CHAPTER OVERVIEW

3.1 Fundamentals of the Brayton Cycle

3.1.1 The Simple Cycle

The operating principle of a gas turbine is based on the Brayton cycle, named after George Brayton who, in the nineteenth century (1874), patented and developed the concept that was originally set forth by John Barber in 1791. The cycle is sometimes referred to James Joule, an English physicist who also presented the concept in 1851.

The cycle is based on two processes at constant entropy – compression and expansion – and two processes at constant pressure – heat addition and rejection. The corresponding thermodynamic diagrams are shown in Figure 3.1.

Cycle work can be evaluated by applying the First Law of Thermodynamics to each component. In doing so, the turbomachines are assumed to be adiabatic (no heat exchange with the surroundings):

Hybrid Systems Based on Solid Oxide Fuel Cells: Modelling and Design, First Edition.
Mario L. Ferrari, Usman M. Damo, Ali Turan, and David Sánchez.
© 2017 John Wiley & Sons Ltd. Published 2017 by John Wiley & Sons Ltd.

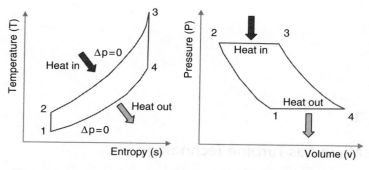

Figure 3.1 Temperature–entropy (left) and pressure–volume (right) thermodynamic diagrams of the Joule-Brayton cycle.

- Compressor:

$$W_c = h_{02} - h_{01} = \bar{c}_{p,c}(T_{02} - T_{01}) \tag{3.1}$$

- Heat adder:

$$Q_a = h_{03} - h_{02} = \bar{c}_{p,ha}(T_{03} - T_{02}) \tag{3.2}$$

- Expander:

$$W_t = h_{03} - h_{04} = \bar{c}_{p,t}(T_{03} - T_{04}) \tag{3.3}$$

- Cooler (heat rejection):

$$Q_r = h_{04} - h_{01} = \bar{c}_{p,hr}(T_{04} - T_{01}) \tag{3.4}$$

where $\bar{c}_{p,x}$ stands for the average isobaric specific heat along process x. Based on these conservation equations, the useful specific work and thermal efficiency are calculated as follows:

$$W_u = W_t - W_c \tag{3.5}$$

$$\eta_B = \frac{W_u}{Q_a} = \frac{Q_a - Q_r}{Q_a} = 1 - \frac{Q_r}{Q_a} \tag{3.6}$$

The p–v diagram in Figure 3.1, where v is the specific volume of the working fluid, is a very useful tool to assess the exchange of mechanical energy in the cycle. To this end, it is enough to consider that the work delivered by a gas in a polytrophic process from 1 to 2 can easily be calculated as:

$$W_1^2 = \int_1^2 v \, dp \tag{3.7}$$

which means that the compression work in a $p - v$ diagram is represented by the area enclosed between the compression line and the line $v = 0$. The same applies for expansion work by merely substituting line $\overline{34}$ for line $\overline{12}$ in Equation 3.7. The combination of equations 3.5 and 3.7 yields the following mathematical expression for cycle specific work:

$$W_u = \oint v\, dp \qquad (3.8)$$

which highlights that the area enclosed by the cycle in a $p - v$ diagram represents the useful work of the cycle.

This is an interesting equation as it provides hints to assess how cycle work can be increased by acting on the following design parameters:

- Inlet conditions: the thermodynamic state of point 1. This is the starting point of the cycle which, for this internal combustion type, corresponds to the environmental pressure and temperature. For ISO reference conditions, this means 15°C and 1.013 bar.
- Compressor delivery pressure: usually defined by specifying compressor pressure ratio.
- Turbine inlet temperature: assuming that pressure remains constant during heat addition, this temperature completely defines the inlet conditions to the turbine.

It is worth noting that there is no need to define further parameters for the heat rejection process, as this takes place at a known pressure (equal to the compressor inlet pressure) and an inlet temperature whose value depends on pressure ratio and turbine inlet temperature. The final temperature of the process is also known.

Let us now take a look at the impact of increasing the pressure ratio by considering Figure 3.2. For the same compressor inlet conditions, compressor delivery pressure is increased from p_2 to $p_{2'}$ whilst turbine inlet temperature is kept constant ($T_3 = T_{3'}$). The area shadowed in light grey indicates the additional cycle work that arises due to the higher expansion ratio across the turbine. The area marked in dark grey represents the loss in cycle work due to a more energy-demanding compression. The impact on useful work results from the balance between these two areas.

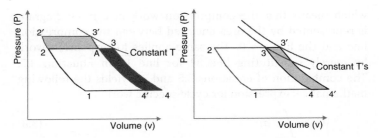

Figure 3.2 Effect of increasing pressure ratio (left) and turbine inlet temperature (right) in a Joule-Brayton cycle.

Increasing turbine inlet temperature increases the area enclosed by the cycle as seen in Figure 3.2 (right). It is a gain in useful work accompanied by a similar increase in heat addition and rejection, but without a detrimental effect in terms of mechanical energy exchange.

Further to this qualitative discussion, a more realistic analysis comes from considering a simplified quasi-ideal Brayton cycle with the following assumptions [1]:

- The working fluid behaves as a perfect gas with constant composition, hence constant isobaric specific heat c_p and specific heat ratio γ, throughout the cycle.
- Constant mass flow rate (i.e. fuel flow rate in a combustion gas turbine is neglected).
- No pressure losses across the heat addition and rejection processes, thus compression and expansion ratios are equal.
- Compression and expansion processes are adiabatic and irreversible, thus being characterized by isentropic efficiencies: η_c and η_t.

With these assumptions, Equation 3.9 can be written as follows:

$$
\begin{aligned}
W_u = W_t - W_c &= (h_{03} - h_{04}) - (h_{02} - h_{01}) \\
&= (h_{03} - h_{04s})\eta_t - \frac{(h_{02s} - h_{01})}{\eta_c} \\
&= c_p T_{01}\left(\left(\frac{T_{03}}{T_{01}} - \frac{T_{04s}}{T_{01}}\right)\eta_t - \left(\frac{T_{02s} - T_{01}}{T_{01}\eta_c}\right)\right)
\end{aligned}
$$

$$= \frac{c_p T_{01}}{\eta_c} \left(\left(\frac{T_{03}}{T_{01}} - \frac{T_{04s}}{T_{03}} \frac{T_{03}}{T_{01}} \right) \eta_t \eta_c - \left(\frac{T_{02s}}{T_{01}} - 1 \right) \right)$$

$$= \frac{c_p T_{01}}{\eta_c} \left(\left(\theta - \frac{\theta}{\delta} \right) \eta_t \eta_c - (\delta - 1) \right)$$

$$W_u = \frac{c_p T_{01}}{\eta_c} (\delta - 1) \left(\frac{\theta}{\delta} \eta_t \eta_c - 1 \right) \tag{3.9}$$

where:

- Pseudo-pressure ratio

$$\delta = \frac{T_{02s}}{T_{01}} = \left(\frac{p_{02}}{p_{01}} \right)^{\frac{\gamma-1}{\gamma}} = \frac{T_{03}}{T_{04s}} = \left(\frac{p_{03}}{p_{04}} \right)^{\frac{\gamma-1}{\gamma}} = \beta^{\frac{\gamma-1}{\gamma}} \tag{3.10}$$

- Temperature ratio

$$\theta = \frac{T_{03}}{T_{01}} \tag{3.11}$$

Total heat addition to the cycle and thermal efficiency are easily calculated by following a similar procedure:

$$Q_a = c_p (T_{03} - T_{02}) = c_p \left((T_{03} - T_{01}) - \frac{(T_{02s} - T_{01})}{\eta_c} \right)$$

$$= \frac{c_p T_{01}}{\eta_c} \left(\left(\frac{T_{03}}{T_{01}} - 1 \right) \eta_c - \left(\frac{T_{02s}}{T_{01}} - 1 \right) \right)$$

$$= \frac{c_p T_{01}}{\eta_c} (\delta - 1) \left(\frac{\theta - 1}{\delta - 1} \eta_c - 1 \right)$$

$$\eta_B = \frac{W_u}{Q_a} = \frac{\frac{\theta}{\delta} \eta_t \eta_c - 1}{\frac{\theta - 1}{\delta - 1} \eta_c - 1} \tag{3.12}$$

These two expressions (3.9 and 3.12) can be used to plot the dependence of useful work and cycle efficiency upon turbine inlet temperature and pressure ratio (θ and δ), shown in Figure 3.3. Several observations are made:

- As expected, specific work is more sensitive to turbine inlet temperature than efficiency, given that the higher expansion work is at the expense of higher heat addition. For instance, increasing the temperature ratio from 3 to 8 has the potential to double the efficiency whilst specific work is multiplied by 7, if an appropriate pressure ratio is selected.

- For a given temperature ratio, the peak efficiency and peak specific work cannot be achieved simultaneously as they take place at different pseudo-pressure ratios. Hence a decision must be made regarding the design pressure ratio of the engine, and whether this will favour size/footprint (work) or fuel economy (efficiency).

The same information plotted in Figure 3.3 can be plotted in a single chart for convenience, as shown in Figure 3.4 where constant θ and δ lines are plotted in solid and dashed lines respectively. The fact that maximum efficiency and specific work cannot be achieved simultaneously for a given θ becomes evident.

Simplifying the quasi-ideal cycle by assuming isentropic compression and expansion has different impacts on specific work and efficiency. Whilst specific work increases slightly, efficiency behaves in a drastically different manner to what was shown in Figure 3.3, exhibiting no apparent dependence on temperature ratio. This is assessed by merely considering $\eta_c = \eta_t = 1$ in equations 3.9 and 3.12:

$$W_{u,i} = c_p T_{01}(\delta - 1)\left(\frac{\theta}{\delta} - 1\right) \tag{3.13}$$

$$\eta_{B,i} = 1 - \frac{1}{\delta} \tag{3.14}$$

where it must be noted, though, that the apparent independence of $\eta_{B,i}$ is illusory, since its value is limited by the efficiency of a Carnot cycle working with similar pressure and temperature ratios:

$$\eta_{C,i} = 1 - \frac{T_{01}}{T_{03}} \tag{3.15}$$

The reason for the notable change in efficiency between the ideal and quasi-ideal cycles is provided in Figure 3.5. Heat addition to the cycle (Q_a) is given by the area enclosed in $(\overline{b23cb})$

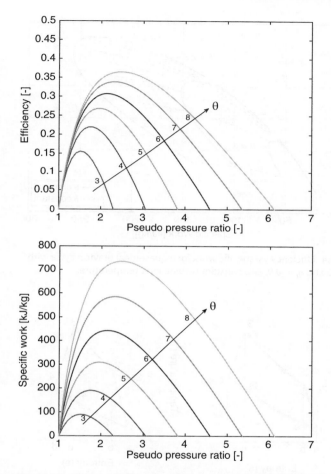

Figure 3.3 Efficiency (left) and specific work (right) for a quasi-ideal Brayton cycle with $\eta_c = 0.85$ and $\eta_t = 0.9$.

whilst heat rejection from it (Q_r) is given by the area enclosed in $(\overline{be4_scb})$.

$$Q_a = \int_2^3 T \, ds \tag{3.16}$$

$$Q_r = \int_1^4 T \, ds \tag{3.17}$$

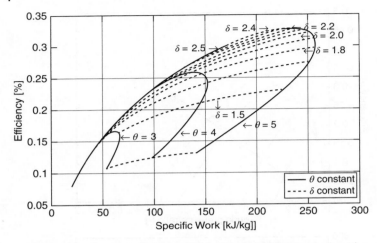

Figure 3.4 Efficiency vs. specific work for a quasi-ideal Brayton cycle with $\eta_c = 0.85$ and $\eta_t = 0.9$, and constant turbine inlet temperature.

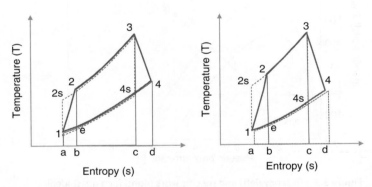

Figure 3.5 Contributions to heat addition and rejection in a quasi-ideal Brayton cycle. Low and high pressure ratio cycles pictured left and right.

Heat rejection can be broken down further into three contributions:

$$Q_r = Q_{r,c} + Q_{r,i} + Q_{r,t} = \int_1^e T \, ds + \int_e^{4s} T \, ds + \int_{4s}^4 T \, ds$$

$$(3.18)$$

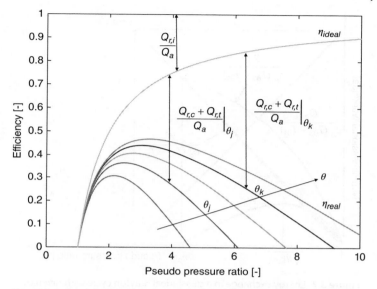

Figure 3.6 Breakdown of losses contributing to the thermal efficiency of a quasi-ideal Brayton cycle.[1]

where $Q_{r,c}$ and $Q_{r,t}$ stand for the heat rejection due to the irreversible performance of the compressor and turbine and $Q_{r,i}$ is the heat rejection from an ideal cycle with the same compressor delivery temperature and pressure ratio.

Based on this breakdown, the main contribution to heat rejection in a cycle with low pressure ratio is $Q_{r,i}$, whilst $Q_{r,c}$ and $Q_{r,t}$ have little impact on efficiency. Nevertheless, when the pressure ratio increases at constant turbine inlet temperature, the fraction of Q_r coming from the irreversible performance of turbomachinery components increases, eventually assuming the dominant influence. This is observed in Figure 3.6 in more detail.

A further analysis of useful work in the quasi-ideal Brayton cycle is helpful to better understand the patterns shown in Figure 3.3. This decomposition of useful work into its constituent terms is illustrated in Figure 3.7 where a rapid rise in turbine work is experienced for increasing pressure ratios in

---

1 Note that η_{ideal} is limited by the efficiency of a Carnot cycle operating with the same temperature ratio, Equation 3.15.

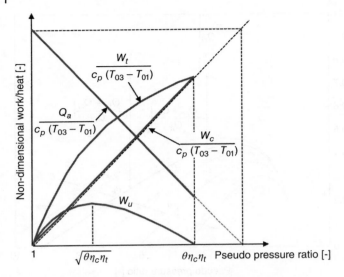

Figure 3.7 Energy exchange in a quasi-ideal Brayton cycle with given θ. *Source*: Haywood (1991) [1]. Reproduced with permission of Elsevier.

the lower δ range; this, in combination with a slower rise in compression work, brings about an increase in useful work.

Peak useful work is found for the value of δ that yields largest separation between W_c and W_t, which turns out to be:

$$\delta_{\eta_{B,max}} = \sqrt{\theta \eta_c \eta_t} \tag{3.19}$$

For pressure ratios higher than this ($\delta > \delta_{\eta_{B,max}}$), compression work increases faster than expansion work and thus useful work decreases. For the entire δ range, heat addition decreases monotonously due to the higher compressor delivery temperature and constant temperature ratio.

3.1.2 The Simple Recuperative Cycle

The previous section set forth the need to find a compromise between fuel economy and engine size. Higher pressure ratios bring about higher efficiencies but, unfortunately, have a very negative impact on specific work. When, in contrast, high specific work is sought, the rise in turbine exhaust temperature brings with it a very large amount of energy delivered to the heat

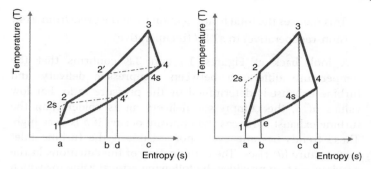

Figure 3.8 Low pressure ratio, recuperative Brayton cycle (left) and high pressure ratio, non-recuperative Brayton cycle (right).

sink (environment). Such is the case for engines operating at a reduced load, regardless of the load control strategy adopted [2]. In these circumstances, it makes sense to try to recuperate this exhaust heat and employ it usefully in the cycle.

The recuperated Brayton cycle is based on exactly this approach (Figure 3.8). It incorporates a heat exchanger (namely a recuperator or regenerator) to transfer sensible heat from turbine exhaust gases (station 4) to compressor delivery air (station 2), hence increasing the temperature of the latter (station 2') and reducing the amount of heat added to the cycle. The higher the difference between T_{02} and T_{04}, the more sense this layout makes.

A closer look at Figure 3.8 reveals that there is a large reduction of both heat addition and heat rejection. Indeed, when the recuperative and non-recuperative cycles are compared for a common temperature ratio:

- Heat addition: the temperature of the working fluid at the inlet to the heat addition process rises from T_2 (compressor delivery) to $T_{2'}$ (high-pressure recuperator outlet). Thus, the area enclosed within figure $\overline{b23c}$ (heat addition to the non-recuperative cycle) is reduced to $\overline{b2'3c}$ (heat addition to the recuperative cycle).
- Heat rejection: the temperature at the inlet to the heat rejection process (flue gases in the real engine) drops from T_4 (turbine exhaust) to $T_{4'}$ (low-pressure recuperator outlet).

This reduces the total heat rejection from the cycle from $\overline{a14d}$ (non-recuperative) to $\overline{a14'd}$ (recuperative).

A look back at Figures 3.5 and 3.8 confirms that the temperature difference between compressor delivery and turbine exhaust is determined by the pressure ratio. For low values of δ, the compressor delivery air is colder than the turbine exhaust, whereas the opposite occurs if δ is very high. These trends become even more visible if the turbine inlet temperature (θ) rises. The combination of the equations in the previous section provides the following critical value for which both temperatures are equal:

$$T_{02} = T_{04} \rightarrow \delta = \sqrt{\theta} \begin{cases} T_{02} > T_{04} \rightarrow \delta > \sqrt{\theta} \\ T_{02} < T_{04} \rightarrow \delta < \sqrt{\theta} \end{cases} \quad (3.20)$$

Let a quasi-ideal recuperative Brayton cycle be considered. This cycle is characterized by the same assumptions as made for the quasi-ideal non-recuperative cycle in the previous section, to which a given efficiency of the heat transfer process in the recuperator is added. This efficiency is defined as the ratio from the temperature rise of compressor delivery air across the heat exchanger ($T_{02'} - T_{02}$) to the maximum attainable temperature rise ($T_{04} - T_{02}$), should the high-pressure air achieve the temperature of the hot turbine exhaust gases at the outlet ($T_{04'} = T_{02}$, $T_{02'} = T_{04}$):

$$R = \frac{T_{02'} - T_{02}}{T_{04} - T_{02}} \quad (3.21)$$

Incorporating the assumption of no pressure losses, there are no changes in the mathematical expression for useful work for given pressure and temperature ratios (Equation 3.9). The total heat added to the standard non-recuperative cycle decreases in an amount equal to the heat recovered from the turbine exhaust gases:

$$\begin{aligned} Q_{a,r} &= c_p(T_{03} - T_{02}) - c_p(T_{02'} - T_{02}) \\ &= c_p(T_{03} - T_{02}) - c_p R(T_{04} - T_{02}) = Q_{a,ref} - Q_{rec} \end{aligned} \quad (3.22)$$

It is reasonable to assume that the temperature difference between compressor delivery and turbine outlet is the same for the quasi-ideal and ideal (isentropic) cycles, for the given temperature and pressure ratios:

$$T_{04} - T_{02} \cong T_{04s} - T_{02s} \tag{3.23}$$

Based on this, the amount of heat recuperated is expressed as follows:

$$Q_{rec} = c_p R(T_{04s} - T_{02s}) = c_p R\left(\frac{T_{03}}{\delta} - T_{02s}\right)$$
$$= c_p R T_{01}\left(\frac{\theta}{\delta} - \delta\right) = c_p R T_{01}\frac{\theta - \delta^2}{\delta} \tag{3.24}$$

The total heat addition to the recuperative cycle is thus:

$$Q_{a,r} = Q_{a,ref} - Q_{rec}$$
$$= \frac{c_p T_{01}}{\eta_c}(\delta - 1)\left(\frac{\theta - 1}{\delta - 1}\eta_c - 1 - \frac{R(\theta - \delta^2)}{\delta(\delta - 1)}\eta_c\right) \tag{3.25}$$

It is evident from this equation that the total heat added to the recuperative cycle is lower than for the non-recuperative counterpart, which, given that specific cycle work is the same for both cases (as long as θ and δ are the same), yields the increased thermal efficiency shown below and plotted in Figure 3.9:

$$\eta_{B,r} = \frac{W_u}{Q_{a,r}} = \frac{\frac{\theta}{\delta}\eta_t\eta_c - 1}{\frac{\theta - 1}{\delta - 1}\eta_c - 1 - \frac{R(\theta - \delta^2)}{\delta(\delta - 1)}\eta_c} \tag{3.26}$$

Again, this information can be plotted in a single chart summarizing the effects of temperature and pressure ratio on specific work and efficiency. This is shown in Figure 3.10 below, where interesting differences and similarities are observed with respect to Figure 3.4. In particular, the recuperative cycle is confirmed to be able to achieve similar specific work at similar pressure ratios in comparison with the non-recuperative cycle, whereas maximum efficiency is now much higher and is achieved for lower pressure ratios. Moreover, it is observed that the constant θ lines are drawn clockwise for increasing pressure ratios in Figure 3.10, whilst it is the opposite in Figure 3.4.

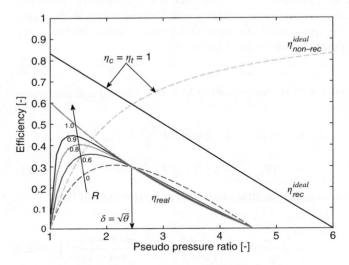

Figure 3.9 Efficiency of the recuperative (solid lines) and non-recuperative (dashed lines) Brayton cycles.

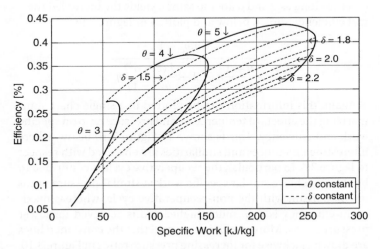

Figure 3.10 Efficiency vs. specific work for a quasi-ideal recuperative Brayton cycle with $\eta_c = 0.85$, $\eta_t = 0.9$ and $R = 0.85$.

The comparisons in Figures 3.9 and 3.10 yield the following conclusions about the quasi-ideal Brayton cycles:

- Incorporating a recuperative heat exchanger does not have any impact on cycle specific work, which essentially depends on the pressure and temperature ratios regardless of the existence of a recuperator.
- Recuperative cycles have the potential to boost efficiency in cycles with low or very low pressure ratios. Moreover, the better the performance of the recuperator, the higher the efficiency attainable and the lower the optimum pressure ratio needed to achieve it.
- The transition pressure ratio beyond which using a recuperative layout is of no further interest corresponds to the pressure ratio for maximum specific work in an ideal Brayton cycle ($\eta_c = \eta_t = 1$), whether recuperative or non-recuperative. In other words, the utilization of a recuperative layout is of no interest if the cycle is designed for lowest footprint/weight.

This summary must nevertheless be appended by some considerations featured by real recuperative cycles, in particular in so far as the first bullet point is concerned:

- One of the main drawbacks of recuperative cycles refers to pressure losses, which imply a reduction of the available enthalpy drop (hence specific work) across the turbine for a given compressor pressure ratio. These pressure losses are found both at the inlet (cold-side pressure loss) and the outlet (hot-side pressure loss).
- Moreover, given that recuperative cycles are typically designed with low pressure ratios, these pressure losses end up reducing an already low specific work. This has a twofold penalty on engine size for a given output.
- For the same reasons, the footprint of the engine is increased twice: (i) due to the very large volume of the recuperator (which has to handle the volumetric flow rate of the engine two times); and (ii) due to the higher throughput coming from a low specific work.

These latter attributes explain why recuperative layouts are not common in large gas turbines, where the footprint of the heat

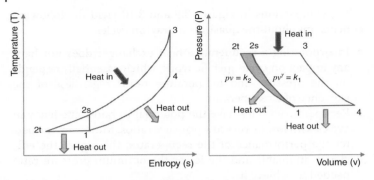

Figure 3.11 Brayton cycle with isothermal compression.

exchanger would become prohibitive in size and cost. In contrast, it is a very interesting option for micro gas turbines where the characteristic mass flow rates and attainable pressure and temperature ratios are low.

3.1.3 The Intercooled and Reheat Brayton Cycles

Intercooling and reheating are means to increase specific work in a Brayton cycle, even though they can potentially reduce efficiency. The cycles adopting any of these layouts are usually termed compound, as opposed to simple cycles where compression and expansion take place in a single adiabatic process without intermediate heat addition/rejection.

Intercooled compression is the practical implementation of the Brayton cycle with isothermal compression whereby compression work is drastically reduced. Such a cycle is shown in Figure 3.11, where process $\overline{12s}$ indicates isentropic compression and $\overline{12t}$ is the compression at constant temperature.

Energy conservation during compression at constant temperature states that:

$$W_c|_T + Q_{r,c}|_T = h_{02t} - h_{01} = 0 \tag{3.27}$$

which implies that an amount of thermal energy equivalent to compression work must be rejected from the machine in order for temperature to remain constant. If this were possible, compression work would be reduced in an amount equivalent to the shaded area in the $p - v$ diagram in Figure 3.11.

Figure 3.12 Ideal Brayton cycle with intercooled compression.

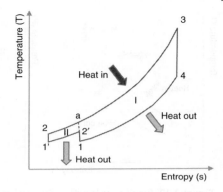

Unfortunately, turbomachines are inherently adiabatic inasmuch as the amount of thermal energy exchanged with the surroundings is negligible with respect to the exchange of mechanical work. This impossibility to compress at constant temperature in a turbocompressor calls for a practical implementation based on a number of adiabatic compression processes in series, with intermediate cooling between them. This is shown in Figure 3.12 where a single intercooler has been considered. It comes naturally that if the number of intercoolers were high enough, compression would eventually take place isothermally.

The efficiency of the ideal intercooled cycle plotted in Figure 3.12 relative to the reference cycle without intercooling is easily assessed merely by considering Equation 3.14. The cycle can be considered to be the addition of two cycles: the original simple cycle with isentropic compression ($\overline{1a341}$) and another cycle with lower pressure and temperature ratios ($\overline{1'2a2'1'}$). Given that the latter has a lower pressure ratio than the former, it also has lower efficiency. As a result, the overall efficiency is lower than in the reference cycle without intercooling.

This rationale cannot be applied when compression and expansion are not considered isentropic, as Equation 3.14 does not hold true anymore, although it might be the case that the intercooled cycle turns out to be more efficient than the non-intercooled layout. It depends on the pressure at which intercooling takes place ($p_{2'}$), the outlet temperature of the

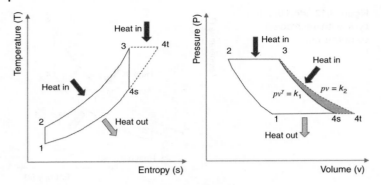

Figure 3.13 Brayton cycle with isothermal expansion.

intercooler $(T_{1'})$ and the isentropic efficiency of the respective turbomachines.[2]

Reheated Brayton cycles are based on a similar approach to increasing specific cycle work, in this case by increasing expansion work at constant compression work. Figure 3.13 shows similar plots to those in Figure 3.11 for the isothermal compression cycle. Again, the higher expansion work is easily visible in the $p - v$ diagram.

Following the same approach as in the intercooled layout, the heat addition during expansion must equal the work produced by the turbine if the gas temperature is to remain constant:

$$W_t|_T + Q_{a,t}|_T = h_{03} - h_{04t} = 0 \tag{3.28}$$

Regarding the efficiency of the isothermal expansion cycle, a negative feature is the extremely high turbine exhaust temperature, which constitutes too heavy a burden to be compensated for by the augmented cycle work; that is, the cycle efficiency decreases in the ideal case.

For similar reasons to those set forth in the Brayton cycle with isothermal compression previously, the practical implementation of the isothermal expansion cycle makes use of adiabatic expansions with intermediate reheat processes that, if numerous enough, resemble the cycle in Figure 3.13. This is shown in

---

2 A more detailed discussion about when the efficiency of this cycle is higher or lower than the cycle without intercooling for the same pressure and temperature ratios is outside the scope of this book.

Figure 3.14 Ideal Brayton cycle with reheat.

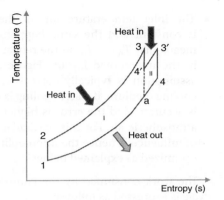

Figure 3.14 for a single reheat cycle. It is observed that the compound cycle comprises the reference cycle with isentropic compression and expansion ($\overline{123a1}$) plus an additional cycle with the same turbine inlet temperature but lower pressure and temperature ratios ($\overline{a4'3'4a}$). The combination of both cycles is of course less efficient than the former, although this statement cannot be extrapolated to the quasi-ideal cycle where the isentropic efficiencies of compressor and turbine are lower than 100%.

Further to the uncertainty regarding cycle efficiency, there is no doubt that compound cycles yield higher specific work. How much higher depends on two main parameters: the intermediate pressure at which reheat/intercooling takes place, and the inlet temperature to the second compression/expansion (i.e. $T_{1'}$ in Figure 3.12 and $T_{3'}$ in Figure 3.14). In order to illustrate this in detail, let the following quasi-ideal cycles be considered:

- The working fluid behaves as a perfect gas with constant composition throughout the cycle, hence constant isobaric specific heat c_p and heat capacity ratio γ.
- Constant mass flow rate, meaning that the fuel flow rate in a combustion gas turbine is neglected.
- No pressure losses across the heat exchangers. Total compression and expansion ratios therefore take the same values.
- Compression and expansion are adiabatic and irreversible, being characterized by a certain isentropic efficiency (η_c and η_t) which is the same for all compressors (η_c) and turbines (η_t).

- The inlet temperature for all the compression processes is constant and the same happens with the turbines. This means that $T_{03'} = T_{03}$ in the reheat cycle, whereas $T_{01'} \approx T_{01}$ in the intercooled layout (Figure 3.15). Whilst the former assumption is typically correct, the latter depends on the cooling medium: if water cooling is used then the assumption is accurate, and the error is higher if the cooling medium is atmospheric air. For this particular case, the only parameter of influence is hence the intercooling pressure, which can be optimized as explained below.

The work consumed by the compressors in the intercooled cycle is expressed as follows:

$$
\begin{aligned}
W_c &= (h_{02'} - h_{01}) + (h_{02} - h_{01'}) \\
&= \frac{c_p}{\eta_c}((T_{02's} - T_{01}) + (T_{02s} - T_{01'})) \\
&= \frac{c_p}{\eta_c}(T_{02's} + T_{02s} - 2T_{01})
\end{aligned}
\tag{3.29}
$$

which means that compression work will be minimum (and hence cycle work will be maximum) if $(T_{02's} + T_{02s})$ is minimum.

$$
\frac{T_{02's}}{T_{01}} = \left(\frac{p_{02'}}{p_{01}}\right)^{\frac{\gamma-1}{\gamma}}
\tag{3.30}
$$

$$
\frac{T_{02s}}{T_{01'}} = \left(\frac{p_{02}}{p_{01'}}\right)^{\frac{\gamma-1}{\gamma}}
\tag{3.31}
$$

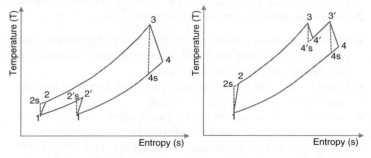

Figure 3.15 Quasi-ideal Brayton cycles with intercooling (left) and reheat (right).

Multiplying these two expressions:

$$\frac{T_{02's}}{T_{01}} \cdot \frac{T_{02s}}{T_{01}} = \left(\frac{p_{02}}{p_{01}}\right)^{\frac{\gamma-1}{\gamma}} \rightarrow T_{02's}T_{02s} = T_{01}^2 \left(\frac{p_{02}}{p_{01}}\right)^{\frac{\gamma-1}{\gamma}} = k$$
(3.32)

The numerical problem is thus to minimize $(T_{02's} + T_{02s})$ knowing that $T_{02's} \cdot T_{02s}$ is constant, which is achieved if $T_{02's} = T_{02s}$. This result implies that:

$$\frac{T_{02's}}{T_{01}} = \frac{T_{02s}}{T_{01}} = \left(\frac{p_{02'}}{p_{01}}\right)^{\frac{\gamma-1}{\gamma}} = \left(\frac{p_{02}}{p_{01'}}\right)^{\frac{\gamma-1}{\gamma}}$$
(3.33)

$$p_{01'}^2 = p_{02'}^2 = p_{01} \cdot p_{02} \rightarrow p_{01'} = p_{02'} = \sqrt{p_{01} \cdot p_{02}}$$
(3.34)

which also implies that $W_{c1} = W_{c2}$.

The application of a similar analysis to the reheat cycle yields the same result; that is, cycle specific work is maximum if:

$$p_{03'}^2 = p_{04'}^2 = p_{03} \cdot p_{04} \rightarrow p_{03'} = p_{04'} = \sqrt{p_{03} \cdot p_{04}}$$
(3.35)

These optimum intermediate pressures for intercooling and reheat are of course only applicable when the aforelisted assumptions are satisfied, which is not necessarily the case in a real engine. Nevertheless, equations 3.34 and 3.35 provide a very good initial estimate for cycle optimization.[3]

3.1.4   The Intercooled and Reheat, Recuperative Brayton Cycle

Intercooling and reheat bring with them a very likely drop in efficiency due to the very large heat addition in the main heater (heavy fuel burn) in the former case, or the very high turbine exhaust temperature in the latter. This is actually the ideal scenario to incorporate reheat in order to enhance efficiency, hence achieving high efficiency and specific work at the same time.

---

3 It must be noted that the simultaneous optimization of intercooling and reheat pressures is only possible when the engine is arranged in a single shaft layout. For multiple shaft arrangements, only one of these pressures can be optimized and the other one results from the power balance in the gas generator shaft(s).

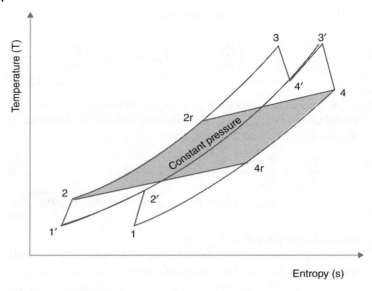

Figure 3.16 Quasi-ideal compound recuperative Brayton cycles with n compression/expansion stages.

Such a cycle would be similar to that depicted in Figure 3.16 where it is noted that intercooling and reheat take place at the same pressure ($p_{02'} = p_{04'}$) as deduced from the previous section.[4] The shaded area in the plot represents the heat exchange in the recuperator.

Using the assumptions already made in the previous section for the quasi-ideal compound cycle, and adding a recuperator with effectiveness R and no pressure losses, mathematical expressions to calculate specific work and efficiency can be derived. These are just a combination of the equations already presented in the previous pages and hence they will not be developed again. Rather, an analysis of the functional dependences will be provided as this is deemed more interesting than a mere theoretical calculation.

---

4 Again, this ideal case with the highest specific work is only possible when the engine is arranged in a single shaft.

The equations providing the specific cycle work and efficiency of a recuperative compound cycle with n compression/expansion stages are:

$$W_u = n \frac{c_p T_{01}}{\eta_c} (\delta^{1/n} - 1) \left(\frac{\theta}{\delta^{1/n}} \eta_t \eta_c - 1 \right) \qquad (3.36)$$

$$\eta_B = \frac{n \left(\frac{\theta}{\delta^{1/n}} \eta_t \eta_c - 1 \right)}{\frac{\theta - 1}{\delta^{1/n} - 1} \eta_c - 1 - \frac{R(\theta - \delta^{2/n})}{\delta^{1/n}(\delta^{1/n} - 1)} \eta_c + (n - 1) \frac{\theta}{\delta^{1/n}} \eta_t \eta_c} \qquad (3.37)$$

Parametric plots of efficiency and specific work are plotted in Figure 3.17 for a compound recuperative cycle with $\theta = 6$ and $R = 0.9$. The following features are worth noting in so far as specific work is concerned:

- It is confirmed that increasing the number of intercooled compressions and reheated expansions brings about a drastic rise in specific work.
- Nevertheless, even if there is a potential to double the specific work by merely going from no intercooling/reheat to one intercooling/reheat, this can only be achieved if a higher pressure ratio is selected. This is similar to selecting a higher temperature ratio in a simple-recuperative cycle (Figure 3.3) but now the associated change in pressure ratio to remain at peak specific work is larger.
- The optimum pseudo-pressure ratio for peak specific work can be calculated as follows:

$$\delta_{W_{max}} = \sqrt[n]{\theta \eta_c \eta_t} \qquad (3.38)$$

- The incremental benefit (for specific work) of an additional compression/expansion stage is reduced when the number of intercooling/reheat stages increases. Actually, for the affordable pressure ratio with state-of-the-art microturbine technology (<3.5), it makes no sense to consider more than one intercooling/reheat stage.

Regarding efficiency, the most interesting change brought about by the compound recuperative layout is found on the right branch of the plots presented in Figure 3.9 for the simple recuperative cycle (i.e. above the pressure ratio for peak

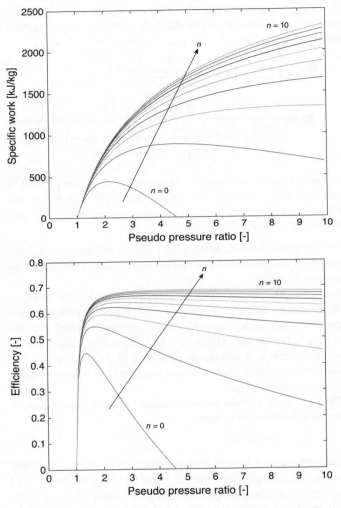

Figure 3.17 Specific work (left) and efficiency (right) of a quasi-ideal compound recuperative Brayton cycle with n compression/expansion stages.

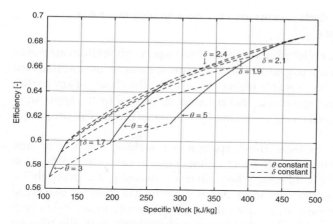

Figure 3.18 Efficiency vs. specific work for a quasi-ideal compound recuperative Brayton cycle with $\eta_c = 0.85$, $\eta_t = 0.9$ and $R = 0.85$.

efficiency). Thus, as opposed to a sudden drop in efficiency, the compound recuperative cycle seems to be much more resistant to increasing pressure ratio. As a matter of fact, if the number of intercooling/reheat stages is increased sufficiently, efficiency becomes virtually insensitive to pressure ratio.

This is a very interesting result for it suppresses the obligatory trade-off between specific work and efficiency. With the layout in Figure 3.16 it is possible to obtain high efficiency and high specific work at the same time, hence offsetting the larger footprint of the heat exchangers with smaller turbomachinery, at least to some extent.

How much the introduction of intercooling/reheat and recuperative heat exchange influences the performance of the Brayton cycle is visualized in Figure 3.18. A quick comparison with Figure 3.10 confirms that the monotonous increase in pressure ratio does not bring about a reduction in specific work. In other words, starting from very low pressure ratios at a given turbine inlet temperature, the constant θ lines always have a positive slope, meaning that they do not achieve a local maximum (at which point the slope would change from positive to negative).

3.1.5 Cycle Layouts used by Contemporary Micro Gas Turbines

All the cycle layouts presented in the previous sections are possible for micro gas turbines, and some of them are used as standard solutions to enable higher efficiencies. The choice of any particular configuration is nevertheless influenced by the application of a specific engine and not by a universal thermodynamic statement.

For instance, those applications seeking lowest cost and footprint typically select a simple non-recuperative cycle, even if this has inherently low efficiency. Actually, it is precisely for this reason why most modern micro gas turbines make use of a recuperative layout due to the difficulty in achieving high pressure (due to some inherent characteristics of the rotating equipment that will be described later) and temperature (micro gas turbines make use of radial expanders that cannot be cooled internally) ratios.

Intercooling is rarely used to reduce compression work when multistage radial turbomachinery is used. This is mainly due to the added pressure losses that significantly affect these low pressure ratio engines (work-wise) and because it reduces the volumetric flow across the second compressor, thus making the corresponding aerodynamic design more challenging. The limited overall pressure ratio is also the main reason why reheat is not used in practice. If it were, the exhaust temperature of the low pressure turbine would be too high for state-of-the-art recuperators, even if turbine inlet temperatures in micro gas turbines are also moderate (ca. 900°C). This is discussed in a later section on recuperators.

In summary, the most widely used cycle layout for micro gas turbines is the simple recuperative cycle without intercooling/reheat. Nevertheless, this is not due to fundamental thermodynamic reasons but for practical limitations. Actually, based on the confirmed interest in the compound recuperative cycle, it is very likely that this will be implemented in the future thanks to progress in materials and manufacturing technologies.

3.2 Turbomachinery

3.2.1 General Considerations on the Selection of Turbomachinery for Micro Gas Turbines

The preliminary design and selection of turbomachinery stages are usually based on dimensional and similarity analyses. As stated in [3], machines that: (i) are geometrically similar; (ii) have similar velocity vectors (diagrams) at similar points in the flow path;[5] (iii) have the same ratio of gravitational to inertial forces acting in the flow path; and (iv) operate with fluids that have the same thermodynamic quality, will have equal fluid dynamic characteristics – that is, equal efficiencies.

This statement can be paraphrased as follows. Those turbomachines that share common values of certain similarity parameters (essentially groups of characteristic parameters of the flow, and operating conditions and geometry of the machine) exhibit the same fluid dynamic performance. The number and definition of these numerical groups is case-specific as discussed in detail in [3] and [4] and will not be discussed here. It is enough to enunciate some of them:

- Head coefficient: representing the enthalpy change (head) across the machine in a pseudo-dimensionless form. It is actually an indirect measurement of pressure ratio:

$$\psi_s = \frac{\Delta h_{0s}}{u_r^2} \tag{3.39}$$

where Δh_{0s} is the isentropic total enthalpy change across the turbomachinery stage and u_r is the blade (peripheral) speed at rotor outlet.

- Stage loading coefficient: this is similar to the head coefficient, although it makes use of the adiabatic enthalpy change. It has

---

5 Balje's statement [3] applies to thermal and hydraulic machinery. For a thermal turbomachine, this condition should be rewritten as "same ratio of inertial to viscous forces", which is actually the same Reynolds number.

a similar physical meaning to ψ_s.

$$\psi = \frac{\Delta h_0}{u_r^2} \tag{3.40}$$

- Flow coefficient: representing the relative size of the machine, that is, the flow passage size relative to the characteristic size:

$$\phi = \frac{c_m}{u_2} \tag{3.41}$$

where c_m stands for the meridional flow speed (or axial velocity in an axial machine).
- Velocity ratio: ratio from the peripheral speed at rotor outlet u_2 to the isentropic velocity c_s, which is:
  - for turbines, the theoretical velocity resulting from an isentropic expansion in a nozzle from the total inlet pressure and temperature to the same backpressure;
  - for compressors, the inlet velocity representing a kinetic energy equal to the static to total isentropic enthalpy rise.
    This parameter is more frequently used in turbines than in compressors.

$$\text{Velocity ratio} = \frac{u_2}{c_s} \tag{3.42}$$

- Degree of reaction: ratio from static enthalpy change across the rotor to total enthalpy change across the stage. In other words, the degree of reaction provides the fraction of the overall static enthalpy change across the stage that takes place in the rotor:

$$R = \frac{\Delta h_{rotor}}{\Delta h_{overall}} \tag{3.43}$$

Making use of the aforementioned stage loading coefficient, it follows that the total head across the turbomachine is proportional to u_r^2 whilst the volume flow rate depends on $C \cdot A \sim u_r \cdot D^2 \sim \omega \cdot D^3$. Thus, if one were to compare the head and volume flow rate of two different turbomachines with similar head and flow coefficients:

$$\frac{\Delta h_0}{\Delta h_{0,ref}} \approx \frac{\omega^2 D^2}{\omega_{ref}^2 D_{ref}^2} \tag{3.44}$$

$$\frac{Q}{Q_{ref}} \approx \frac{\omega D^3}{\omega_{ref} D_{ref}^3} \qquad (3.45)$$

These equations can be solved for rotating speed and diameter by merely assuming that the reference values of Δh_{0s} and Q are unity. Rearranging the results, the following expressions are obtained:

$$\omega_{ref} = \frac{\omega \sqrt{Q}}{\Delta h_0^{3/4}} = \omega_s \qquad (3.46)$$

$$D_{ref} = \frac{D \, \Delta h_0^{1/4}}{\sqrt{Q}} = D_s \qquad (3.47)$$

Based on these considerations, ω_{ref} and D_{ref} are respectively the rotating speed and diameter of a turbomachine handling a volume flow rate of unity across a unit head (whether expanding or compressing). These parameters are termed specific speed and specific diameter, namely ω_s and D_s, and they enable the statement of the following: *'turbomachines of similar geometry (i.e. radial or axial) and specific parameters (ω_s/D_s) have similar flow mechanisms and, therefore, achieve similar efficiencies'*. This statement holds true as long as the effects of Mach and Reynolds numbers are negligible [5].

A well-established relation is hence set between the specific parameters of a turbomachine and the most appropriate type of equipment for a particular duty. This is shown graphically in Figure 3.19 [6] which relates specific speed to machine type and expected efficiency. For the optimum efficiency, another set of figures enables the selection of the corresponding specific diameter, depending upon which the efficiency might differ from the maximum affordable value in Figure 3.20. (ω_s, D_s) pairs on the Cordier line provide highest efficiency, whilst this parameter (η) drops when the design departs from the line.

The specific work of an ideal simple non-recuperative Brayton cycle with 900°C turbine inlet temperature and 3:1 pressure ratio working in ISO conditions is of the order of 150 kJ/kg, although it might vary largely in a real application, depending on turbomachinery efficiency. This means that, considering microturbines in the range from 5–250 kWe, the expected mass flow rates would take values between 0.03 and 1.5 kg/s. Elementary

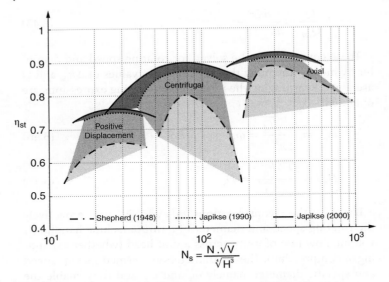

Figure 3.19 Selection of turbomachinery type and attainable total to static efficiency based on specific speed. *Source*: Whitfield and Baines (1990) [6]. Reproduced with permission of Elsevier.

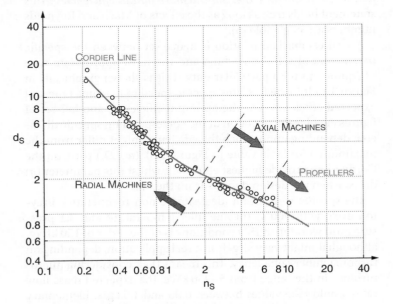

Figure 3.20 Specific speed–specific diameter pairs for peak compressor efficiency (Cordier line).

calculations for these operating conditions yield specific speeds between 0.5 and 1.5 for the compressor, which lie well within the range of radial turbomachinery in Figures 3.19 and 3.20. Based on this, compressor efficiencies in the range of 70–80% are to be expected. Similar calculations for the expander recommend radial inflow turbines.

This approach is quite common to obtain a first selection of the most suitable type of turbomachinery for a micro gas turbine, as well as a preliminary estimate for efficiency from which further cycle calculations can be performed. Then, a more refined analytical/experimental design process follows for optimization and component integration issues. To this end, any of the numerous handbooks on the topic can be used.

3.2.2 Fundamentals of Radial Compressor Design and Performance

Turbomachinery performance is based on two fundamental sets of equations: the Euler equation and the energy balance equations. The Euler equation is also known as the *fundamental turbomachinery equation* and allows us to estimate the energy (work) exchange in a turbomachinery stage with minimum information. Fluid flow in a turbomachinery stage is complex and difficult to characterize (three-dimensional and unsteady), especially when radial machines are concerned. This poses challenges when calculating the work needed to overcome a certain pressure ratio or the work that can be obtained from a given available pressure drop. Nevertheless, the application of the law of conservation of angular momentum contributes to carrying out such a task rationally. After manipulation, this yields the following expression:

$$W_{st} = u_2 c_{\theta 2} - u_1 c_{\theta 1} \tag{3.48}$$

where $u_{1,2}$ and $c_{\theta 1,2}$ stand for the blade speed (or peripheral velocity) and whirl velocity at impeller inlet (1)/outlet (2) respectively (see Figure 3.21).

Equation 3.48 is formally applied to an individual streamline since, in reality, the flow velocities are not uniform at the inlet and outlet sections of the impeller. This is mainly due to the span-wise variation of peripheral velocity at the inlet (u_1) and

Figure 3.21 Velocity diagrams in a radial compressor impeller.

the tangential variation of whirl velocity at the outlet ($c_{\theta 2}$). In order to extrapolate the equation to the whole stage, it must be assumed that the flow velocities are uniform at the inlet and outlet sections of the impeller. This is called the unidimensional approach to radial compressor analysis which, owing to the neglected non-uniformities of the flow, overestimates the amount of work exchanged between impeller and fluid.

The amount of work exchanged can be visually displayed in the enthalpy vs. entropy diagram of the stage merely by applying the following conservation equations at the impeller inlet (1) and outlet (2) and at the outlet of the diffuser (3), on the assumption that the stage is adiabatic but not isentropic. The notation is easily interpreted in Figure 3.22.

$$h_{03} - h_{01} = W_{st} = u_2 c_{\theta 2} - u_1 c_{\theta 1} \tag{3.49}$$

$$h_{02r} - h_{01r} = \frac{u_2^2 - u_1^2}{2} \tag{3.50}$$

$$h_{03} - h_{02} = 0 \tag{3.51}$$

The efficiency of the compressor stage is defined with the h–s diagram in Figure 3.22. For any mechanical compression device, efficiency is commonly defined as the ratio from minimum to actual compression work necessary to overcome the same pressure difference between inlet and outlet.

$$\eta_c = \frac{W_{st,min}}{W_{st,actual}} \tag{3.52}$$

Figure 3.22 Enthalpy vs. entropy diagram of the radial compressor stage.

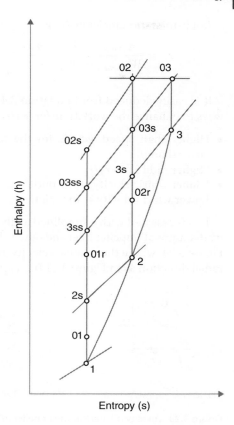

Two definitions of minimum compression work are possible:

- Isentropic work needed to achieve the same static pressure (p_3) and velocity (c_3) at the outlet. This yields to the 'total-to-total' efficiency (η_{tt}):

$$\eta_{tt} = \frac{W_{st,min}}{W_{st,actual}} = \frac{h_{03ss} - h_{01}}{h_{03} - h_{01}} \tag{3.53}$$

- Isentropic work needed to achieve the same static pressure (p_3) at the outlet and, ideally, no kinetic energy ($c_3 = 0$) since this is considered a loss. The resulting efficiency is the 'total-to-static' efficiency (η_{ts}):

$$\eta_{ts} = \frac{W_{st,min}}{W_{st,actual}} = \frac{h_{3ss} - h_{01}}{h_{03} - h_{01}} \tag{3.54}$$

It is easily deduced from Equation 3.49 that there are several ways to enhance the work transfer across the impeller:

- Higher shaft speed, which, for the same radii, increases u_2 faster than u_1.
- Higher radii ratio r_2/r_1.
- Higher whirl velocity at the outlet, $c_{\theta 2}$.
- Lower whirl velocity at the inlet, $c_{\theta 1}$.

This is assessed with a simplified analysis of the generic velocity diagrams at impeller inlet and outlet shown in Figure 3.23 and Figure 3.24, where the relative velocity at the outlet is not in the radial direction as in Figure 3.21 (i.e. $c_{\theta 2} \neq 0$).

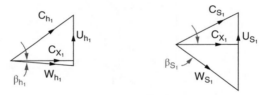

VELOCITY DIAGRAM AT HUB   VELOCITY DIAGRAM AT SHROUD

Figure 3.23 Velocity diagram at the impeller inlet (hub and shroud) for pre-whirl inflow.

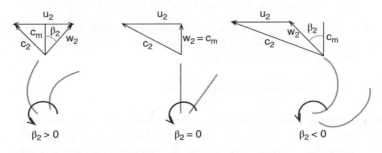

Figure 3.24 Velocity diagram at impeller outlet for different outlet flow (sweep) angles.

a) *Inlet flow*: the absolute velocity of the inlet flow to the impeller c_1 is usually in the axial direction and has a constant value spanwise. Accordingly, Euler's equation has the following default expression for the attainable work transfer:

$$W_{st} = u_2 c_{\theta 2} - u_1 c_{\theta 1} = [c_{\theta 1} = 0] = u_2 c_{\theta 2} \qquad (3.55)$$

The utilization of inlet guide vanes can nevertheless alter this situation, providing the flow with a variable inlet angle (also variable spanwise) in order to either increase or reduce the absolute whirl velocity at the inlet ($c_{\theta 1} \neq 0$). Two situations are possible:

- Pre-whirl: positive inlet flow angle ($\alpha_1 > 0$) bringing about a certain velocity in the same direction as blade speed ($c_{\theta 1} > 0$) with the following effects:
    - For the same shaft speed and impeller geometry at the outlet, lower stage work and pressure ratio.
    - Larger area at impeller (inducer) inlet. This is so regardless of whether the axial velocity remains constant. If it does, then the static pressure drops and so does density. If it does not, then density remains constant but axial velocity decreases. Such an area increase might be beneficial in applications requiring very small sizes.
- Counter-whirl: negative flow angle ($\alpha_1 < 0$) bringing about a certain velocity that opposes the blade speed ($c_{\theta 1} < 0$) with the following consequences:
    - For the same shaft speed and impeller geometry at the outlet, higher stage work and pressure ratio.
    - Same effect on the inlet flow area.

For operational and manufacturing reasons, axial inlet is the design of choice in most applications. Nevertheless, pre-whirl is sometimes employed to decrease the relative Mach number at the inlet, hence avoiding severe shock-induced boundary layer losses. This enables higher rotating speeds and pressure ratios (it must be noted that the increase of $u_2 c_{\theta 2}$ relative to $u_1 c_{\theta 1}$ when ω increases is much higher). For similar reasons, counter-whirl is seldom used as it increases the relative tip Mach number at the inlet, thus deteriorating stage performance [4].

b) *Outlet flow*: relative flow angle at impeller outlet is typically referred to as sweep angle. Radial blades have a twofold benefit, namely reduced cost (ease of manufacturing) and absence of bending moment due to centrifugal forces. Still, this angle is commonly modified with respect to the radial direction for a number of reasons. Three design options are possible:

- Radial blades: most common layout for the cited reasons. Stage reaction is close to 0.5, meaning that pressure rise is evenly distributed between impeller and diffuser.
- Backswept blades: relative velocity faces backwards in the sense of rotation ($\beta_2 > 0$). This means a lower absolute velocity at impeller outlet (lower $c_{\theta 2}$) for the same impeller diameter and shaft speed, which reduces the stage work and pressure ratio. Stage reaction increases and so does efficiency.[6]
- Forward-swept blades: relative velocity does not increase significantly with respect to the previous cases but, given that it is now in the direction of rotation ($\beta_2 < 0$), a very large increase of absolute velocity at impeller outlet is experienced. Accordingly, the stage work and pressure ratio increase whilst stage reaction decreases. Efficiency is typically reduced as well, not only in rated conditions but also in off-design operation due to the higher sensitivity to the incidence angle of the diffuser. This configuration presents severe stability issues, especially at off-design conditions, and it is not usually employed due to surge and other aerodynamic and mechanical integrity considerations.

These features are summarized in Table 3.1, showing the influence of sweep angle on compressor performance.

The utilization of non-dimensional parameters shown at the beginning of this section in turbomachinery provides a quick assessment of the expected performance of the stage for varying sweep angles. Indeed, for a centrifugal compressor stage with axial inlet velocity ($c_{\theta 1} = 0$) handling a perfect gas,

---

6 Stage efficiency benefits from the better performance of the impeller compared with the diffuser (higher total-to-total efficiency) and the lower outlet velocity from this latter component (higher total-to-static efficiency).

Table 3.1 Influence of sweep angle on stage performance.

	Backswept	Forward-swept
Work	↓	↑
Pressure ratio	↓	↑
Reaction	↑	↓
Efficiency	↑	↓

the performance of the stage can be expressed as follows:

$$\frac{p_{03}}{p_{01}} = \left[1 + (\gamma - 1)\eta_{tt}\left(\frac{u_2}{a_{01}}\right)^2(1 - \phi_2 \tan(\beta_2))\right]^{\frac{\gamma}{\gamma-1}} \quad (3.56)$$

$$R = \frac{1}{2}(1 + \phi_2{}^2 \tan(\beta_2)) \quad (3.57)$$

where:

- γ: ratio of specific heats.
- η_{tt}: total-to-total efficiency of the stage.
- u_2: blade speed at impeller outlet.
- a_{01}: sound speed at compressor inlet, referred to total temperature ($\sqrt{\gamma R_g T_{01}}$).
- $\frac{u_2}{a_{01}}$: blade Mach number.
- ϕ_2: flow coefficient.
- β_2: blade angle at impeller outlet.
- R: stage degree of reaction.

Equations 3.56 and 3.57 provide an excellent tool to assess the expected performance of a compressor stage when the design parameters are changed, as they summarize the foregoing discussion regarding the rotational speed, outlet radius and sweep angle. The same information can be presented graphically as in Figure 3.25, where the dashed and solid lines show the expected efficiency under ISO conditions for 500 m/s and 625 m/s peripheral speed (u_2). The effect of increasing total-to-total efficiency is to shift the constant (β_2) lines upwards, whilst, on the other hand, increasing the flow coefficient reduces the pressure ratio for a given blade Mach number (i.e. it reduces $c_{\theta 2}$ owing to the higher meridional velocity c_m).

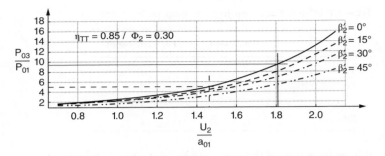

Figure 3.25 Dependence of pressure ratio on blade Mach number and outlet blade angle.

Regarding stage reaction, this is higher for higher blade outlet angles, all other parameters being equal. This is easy to understand if the absolute velocity at impeller outlet is monitored. Any non-radial configuration of the impeller blades brings about an increase in relative velocity (w_2), thus decreasing the relative flow diffusion, as shown in Figure 3.24. At the same time, bending the blades backwards reduces absolute velocity (c_2) which, for the same diffuser geometry (r_3/r_2), implies a lower enthalpy rise across this component. The contrary occurs when forward-swept impellers are used.

The net effect on stage reaction is easily assessed if the following equation is used:

$$R = \frac{(w_1^2 - w_2^2) + (u_2^2 - u_1^2)}{(w_1^2 - w_2^2) + (u_2^2 - u_1^2) + (c_2^2 - c_1^2)} \tag{3.58}$$

The dominant effect of $(c_2^2 - c_1^2)$ on stage reaction implies that increasing β_2 increases the fraction of work used to increase enthalpy, whilst a lower fraction of this is employed to increase total kinetic energy across the impeller. The contrary also happens; when forward-swept blades are used, emphasis is put on increasing total kinetic energy, whilst a lower fraction of stage work is dedicated to raising the enthalpy of the fluid.

The impact of sweep angle on absolute velocity (c_2) is actually very important in determining the operating range of the compressor due to its influence on the performance characteristics of the diffuser. Unlike axial-type machines, radial compressors

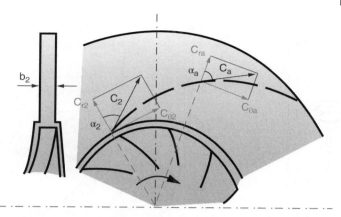

Figure 3.26 Trajectory (streamline) of a fluid particle in a vaneless diffuser.

do not necessarily have blades in the stator. Indeed, simple parallel plates downstream of the impeller serve to reduce flow velocity merely by increasing the cross-sectional area. For such geometry, a fluid particle entering the diffuser would follow a curved trajectory based on continuity (controlling c_r) and conservation of angular momentum (controlling c_θ) considerations (see Figure 3.26):

$$\tan \alpha_a = \frac{c_{\theta a}}{c_{ra}} = \frac{\rho_a c_{\theta 2}}{\rho_2 c_{r2}} = k \frac{\rho_a}{\rho_2} \tag{3.59}$$

The length of the trajectory followed by particle a in Figure 3.26 determines the diffusion ratio (enthalpy rise) and friction losses. This is shown in Figure 3.27, where lines of constant pressure rise are plotted for different radii ratios (r_3/r_2) and height ratios (b_3/b_2).[7] It is easily observed that the only way to increase the diffusion ratio is either to increase the diffuser outlet radius (r_3) or the height (b_3), or to employ a vaned diffuser. This is further confirmed by the studies of Bradshaw and Brown which show that, for $r_3/r_2 \in [1.8, 3.0]$, there is no real improvement in compressor efficiency due to the augmented friction losses [7].

The theoretical length travelled by a particle entering a vaneless diffuser with $\alpha_2 = 70°$ in a diffuser with $r_3/r_2 = 1.8$

---

7 *b* stands for diffuser height.

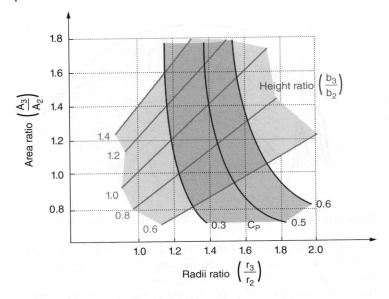

Figure 3.27 Ideal pressure rise coefficient $C_{p,i}$ of a vaneless diffuser (diffuser inlet angle $\alpha_2 = 70°$).

corresponds to a complete turn of a logarithmic spiral (360°). This very long path brings about very high friction losses and thus poor static pressure rise ($C_p \approx 0.5$). Nonetheless, a vaned diffuser enables longer trajectories for the same cross-sectional area change, thus yielding higher diffusion ratios for the same radii ratio or, conversely, lower compressor size for given diffusion ratio, even if this is at the expense of a higher manufacturing cost. There are two main design parameters in a diffuser: number and shape of blades. The number of blades comes determined by two main criteria:

- Expected operating range: a higher number of blades enables better flow guidance, hence yielding higher compressor efficiency for low specific speed applications. Nevertheless, if the mass flow rate increases during operation, incidence losses rise rapidly as a consequence of the variations of α_2. When added to the also high frictional and blockage losses, this brings about a visible drop in efficiency. If, on the other

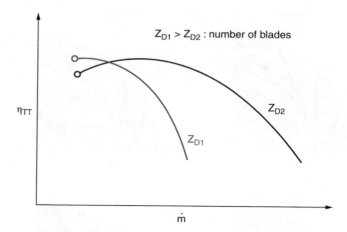

Figure 3.28 Impact of the number of diffuser vanes on the total-to-total efficiency and range of a centrifugal compressor.

hand, fewer blades are used, then guidance is insufficient at low flow rates, having a negative impact on efficiency, as shown in Figure 3.28. As a final note, a high number of blades also reduces the surge margin[8] of the compressor.

- Number of impeller blades: the numbers of blades in the impeller and diffuser must be odd, as otherwise resonance and aeroelastic coupling problems could arise. Also, there must be fewer diffuser vanes than impeller blades, in order to avoid uneven pressure rise in adjacent diffuser channels that could eventually bring about backflow or choke issues due to strong secondary flows at the impeller outlet [8]. Regarding the number of impeller blades, the following formula by Pfleiderer provides a first estimate [9]:

$$Z_I = 6.5 \frac{\left(r_2/r_1 \right) + 1}{\left(r_2/r_1 \right) - 1} \cos \frac{\beta_1 + \beta_2}{2} \tag{3.60}$$

It must be noted, though, that this formula was originally developed for pumps, so even if the fundamental flow physics of fluid compression are the same, the estimate is just approximate.

---

8 Surge corresponds to the instable operation of the compressor. This is described further in a dedicated section.

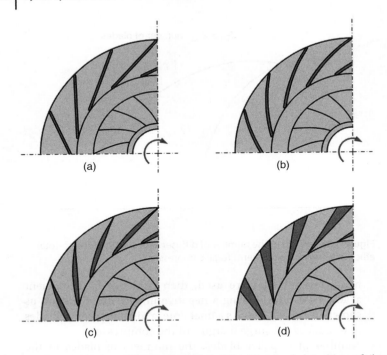

Figure 3.29 Various diffuser types: (a) straight, (b) curved plate, (c) aerofoil, (d) wedge.

As easily deduced, there is a close link between impeller and diffuser design. The integration of the foregoing discussion yields the following main conclusions regarding stage design, in particular sweep angle (β_2) and absolute flow angle (α_2) at impeller outlet:

- High backswept impellers are typically used in applications looking for extended operating range and moderate pressure ratios (<3), in combination with vaneless diffusers. According to [8], there is a 1–2 percentage point efficiency rise and a 25% range extension for each 10° backsweep angle increase.
- For absolute flow angles higher than $\alpha_2 = 70°$, there is a drastic drop in diffuser efficiency that should be corrected with a

vaned diffuser.[9] In this latter case, higher inlet angles to the diffuser are admissible ($70° < \alpha_2 < 75°$).

- Wedge (channel) diffusers provide highest pressure ratios but also a lower operating range, whilst the performance of an aerofoil (cascade) diffuser lies somewhere between wedge and vaneless diffusers. The most usual types of diffuser vanes in centrifugal compressors are shown in Figure 3.29.

3.2.3 Some Notes on Compressor Surge

'Surge' refers to the unstable operation exhibited by a turbocompressor when forced to act at a very high pressure ratio and low volumetric flow rate. It is therefore a general instability which cannot be mistaken for stall, which shares some common features but takes place locally inside the machine.

Figure 3.30 shows the performance map of a turbocompressor operating at constant speed in point A. Closing the discharge throttle valve partially brings about a higher pressure ratio and, subsequently, lower flow rate, until point 1 is reached. Should the discharge valve be closed further beyond this point, the compressor running at constant speed would not be able to overcome the pressure difference between the suction and discharge sections, and thus flow reversal (gas flowing from discharge to suction) would arise. This would ideally be represented by point 2 in the plot.

Reverse flow rate is proportional to the shaft speed of the compressor and, especially, the characteristics of the discharge volume, and has the main impact of relieving the high pressure in the compressor discharge volume, thus reducing the negative volume flow rate; this is represented by line $\overline{23}$ in the plot. When the discharge pressure is reduced to P_3, the compressor is able to overcome the pressure gradient between suction and discharge and hence a forward flow is established again [10]. This is point 4 in Figure 3.30. From this point on, discharge pressure starts to build up again and, if no preventive actions are taken, point 1 is reached and the loop starts again [11]. But this is a theoretical description of surge and, in practice, flow reversal does not usually take place. Instead, an alternating variation of volume

---

9 Japikse and other authors make use of the swirl angle $\lambda_2 = c_{\theta2}/c_{r2} = \tan \alpha_2$, which is obviously equivalent to α_2 [8].

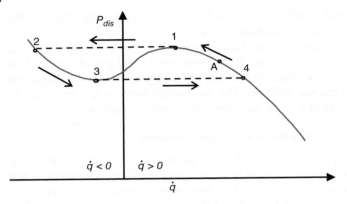

Figure 3.30 Theoretical surge loop in a turbocompressor.

flow rate around the flow through the throttle valve, and a corresponding pressure build-up and decay, is observed [12].

Point 1 would theoretically be known as the surge point, defined as the point of peak head capability and minimum flow limit [13]. Any operating point to the left of the surge point lies within the unstable region of the compressor, whilst the operation far rightwards is safe. The stability of operating the compressor to the right but near the surge point depends largely on the dynamics of the system, and thus it cannot be ensured that the surge point is always found at the highest pressure ratio for a given shaft speed. Furthermore, the surge point in an axial compressor is typically to the right of the theoretical peak head, whilst the contrary occurs in a centrifugal compressor, for which the point of peak pressure ratio is stable. The line linking all the surge points of the compressor at different speeds is the surge line and determines the region of unstable operation.

This information is shown in Figure 3.31 depicting the performance map of a centrifugal compressor. The map is made up of the following lines:

- Constant speed lines (solid): these lines provide the pressure ratio attained by the compressor at a given mass flow rate and rotating speed.
- Efficiency islands (short-dashed): these are contours of constant total-to-total efficiency. It is observed that peak efficiency typically lies well within the stable region.

- Surge line (long-dashed): line separating the stable and unstable operating regions. Operating conditions that lie on the left of this line are not feasible.
- Running line (dash-dot): the control system ensures that the operating conditions of the compressor lie on this line in order to achieve the highest efficiency for each mass flow rate or pressure ratio. This can be achieved by various means such as changing the shaft speed and/or acting on a recirculation loop to increase the mass flow rate through the compressor. The star on the plot indicates the design point (DP), which corresponds to the rated operating conditions of the machine.

The distance between the running and surge lines in the plot is the surge margin:

$$SM = \left. \frac{\dot{m}_{op} - \dot{m}_{sg}}{\dot{m}_{op}} \right|_{N_{cor}} \tag{3.61}$$

where $\dot{m}_{op}$ and $\dot{m}_{sg}$ stand for the operating mass flow rate and the mass flow rate at which surge occurs for a constant corrected

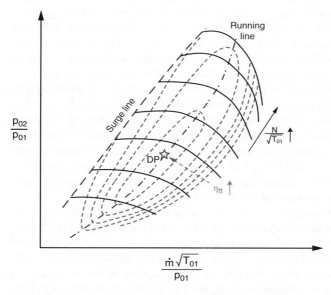

Figure 3.31 Performance map of a centrifugal compressor.

shaft speed (N_{cor}). The magnitude of the surge margin depends on the characteristics of the machine (smaller for axial compressors, larger for centrifugal machinery), the expected operation (whether steady or subjected to frequent transients) and, very importantly, the characteristics of the system downstream of the compressor. As a rough value, it could be established that *SM* should be higher than 10–15%.

Surge has the potential to cause severe damage to the rotating parts of the compressor and in the seals and bearings [13]. It is therefore necessary to implement surge-avoidance measures in applications comprising turbocompressors, as is the case for micro gas turbines. These can be of different types, but they all share a common purpose – to keep the surge margin above a lower limit:

- *Discharge (blow-off) valves*: one or more fast-acting valves are located downstream of the compressor. When there is a sudden drop in mass flow rate through the compressor, for instance due to a rapid reduction in shaft speed, its head capacity falls whilst the discharge pressure remains initially constant (as it is commanded by the downstream system). These operating conditions would immediately run the compressor into surge and thus a set of fast-acting blow-off valves are responsible for relieving the discharge pressure in the same timescale that the mass flow rate is decreasing.
- *Recycle valves*: when the operating conditions that are likely to run the compressor into surge build up more slowly, a recycle valve is used to increase the mass flow rate through the compressor with respect to the circulating mass flow rate through the system. This system is standard in gas compression systems, though it is less interesting for a micro gas turbine due to the increased compression power that can potentially lead to a mechanical work imbalance.
- *Variable geometry*: beyond stage design, the utilization of variable geometry compressors enables controlling (i) inlet relative Mach number at the tip (via inlet guide vanes, IGV) and (ii) incidence losses at diffuser inlet (via variable geometry diffusers). These changes in the geometry of the system yield a new compressor map, thus a relocation of the surge line, and enable continuous control of the surge margin over a very

wide operating range. This is shown in Figure 3.32, where the impact of variable geometry on the compressor map can be seen [14–15]. Attention is drawn to the extremely wide operating range of the compressor incorporating variable geometry.

These three are the most common surge-avoidance measures used in different applications, and they can also be combined, even with other systems not listed here. It is important to note that there are differences between them, both in the reaction time and the impact that they have on overall system performance. Thus blow-off valves are fast-acting, and although they can be used for continuous control purposes (bleed valves), they are mostly used for emergency shutdown operation, given that they increase compression power unnecessarily. Recirculation valves are slower and not normally used in an emergency. Their use is common in gas pumping stations where they are very useful to bring the compressor on/off-line in a controlled manner [10, 13]. Again, a large recirculation ratio increases compression power. Variable geometry is most desired in the sense that it does not bring about higher compression work but, on the contrary, keeps this to a minimum.

In the light of the considerations above, it is easily deduced that the optimum surge-avoidance scheme for a micro gas turbine would emerge from the combination of a blow-off valve, able to immediately relieve the discharge pressure of the compressor in case of an emergency shutdown, and variable geometry: variable inlet guide vanes and, ideally, diffuser vanes. In the case that variable geometry is not implemented, then a bleed-valve in parallel to the blow-off valve (rather than the blow-off valve itself) would enable continuous surge control [16]. Other more complex surge-avoidance schemes are also possible, as shown in [17] for a fuel cell and micro gas turbine hybrid system.

3.2.4 Fundamentals of Radial Turbine Design and Performance

The fundamentals of a radial inflow turbine are very similar to those of a centrifugal compressor stage. A set of stator vanes expands the fluid, increasing the absolute velocity and reducing the static pressure and temperature. This high kinetic

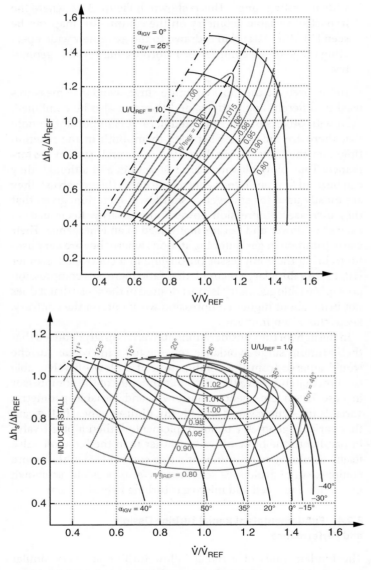

Figure 3.32 Performance map of centrifugal compressor stage: reference fixed geometry unit (top) and unit incorporating variable inlet guide vanes and diffuser vanes (bottom). Adapted from [14] with permission.

energy, along with a fraction of the remaining pressure energy (enthalpy) at wheel inlet, is then converted into useful shaft work. The turbine is usually designed with a centripetal flow in order to take advantage of a positive work developed by centrifugal forces.[10]

The amount of work exchanged can be correlated to the enthalpy vs. entropy diagram of the stage by applying the following conservation equations at the stator inlet (1) and the inlet (2) and outlet from the wheel (3), on the assumption that the stage is adiabatic but not isentropic. The notation is easily

Figure 3.33 Enthalpy vs. entropy diagram of the radial inflow turbine stage.

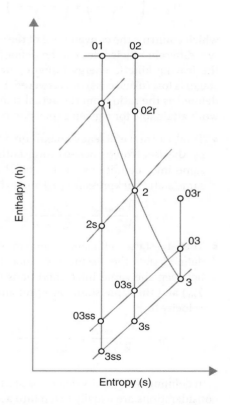

interpreted in Figure 3.33.

$$h_{01} - h_{02} = 0 = (h_1 - h_2) + \left(\frac{c_1^2 - c_2^2}{2} \right) \tag{3.62}$$

$$h_{02} - h_{03} = W_{st} = u_2 c_{\theta2} + u_3 c_{\theta3} \tag{3.63}$$

$$h_{02r} - h_{03r} = \frac{u_2^2 - u_3^2}{2} = (h_2 - h_3) + \left(\frac{w_2^2 - w_3^2}{2} \right) \tag{3.64}$$

If Equations 3.63 and 3.64 are combined, the following expression is obtained:

$$W_{st} = \left(\frac{c_2^2 - c_3^2}{2} \right) + \left(\frac{u_2^2 - u_3^2}{2} \right) + \left(\frac{w_3^2 - w_2^2}{2} \right) \tag{3.65}$$

which confirms the convenience of the centripetal layout. Again, two different efficiencies can be defined depending on whether the leaving kinetic energy (kinetic energy at the outlet of the stage) is lost (usual case) or recovered. In both cases, efficiency is defined as the ratio from the actual stage work to the maximum work attainable for a given expansion ratio.

- Total-to-total efficiency: maximum work attainable is defined by the isentropic total-to-total enthalpy drop between the same inlet conditions as the actual stage (p_{01}, T_{01}) and the same static backpressure (p_3) and exhaust velocity (c_3):

$$\eta_{tt} = \frac{W_{st,actual}}{W_{st,max}} = \frac{h_{01} - h_{03}}{h_{01} - h_{03ss}} \tag{3.66}$$

- Total-to-static efficiency: maximum work attainable is defined by the isentropic total-to-total enthalpy drop between the same inlet conditions as the actual stage (p_{01}, T_{01}) and the same static backpressure (p_3), with null exhaust velocity (c_3):

$$\eta_{ts} = \frac{W_{st,actual}}{W_{st,max}} = \frac{h_{01} - h_{03}}{h_{01} - h_{03ss} + c_3^2/2} = \frac{h_{01} - h_{03}}{h_{01} - h_{3ss}} \tag{3.67}$$

In defining the velocity diagrams of a radial inflow turbine, two considerations are usually taken into account:

- The rotor blades are radial at the inlet, which is the standard radial inflow design approach.

- A certain negative incidence ranging from $-20°$ to $-40°$ is usually adopted in order to compensate for the very high aerodynamic load of the blades. Given the radial layout of the blades, this incidence angle is in practice β_2.
- In order to maximize total-to-static efficiency, the rotor is designed for axial discharge velocity c_3 ($\alpha_3 = 0$).

These considerations yield the velocity diagrams in Figure 3.34 and the following expression for Euler's equation, where the same considerations about flow uniformity set forth for the compressor must be applied (i.e. this equation is only valid if a unidimensional approach is adopted):

$$W_{st} = u_2 c_{\theta 2} + u_3 c_{\theta 3} = u_2 c_{\theta 2} \tag{3.68}$$

A very useful tool to assess the number of stages for a given available enthalpy drop, or alternatively to define the optimum blade speed at rotor inlet, is the isentropic velocity c_s. This velocity, which is usually referred to as spouting velocity when applied to radial turbine design, is the theoretical flow velocity if the working fluid were expanded in an isentropic nozzle from the total inlet pressure and temperature of the turbine to its same backpressure:[11]

$$\frac{1}{2}c_s^2 = h_{01} - h_{3ss} \tag{3.69}$$

It is observed that, for an impeller with radial relative inflow ($\beta_2 = 0°$) and axial absolute outflow (no outlet swirl $c_{\theta 3} = 0$),

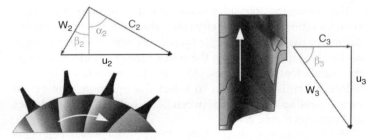

Figure 3.34 Velocity diagrams of a radial inflow turbine.

---

11 Depending on the designer's interest, total exhaust pressure or static exhaust pressure can be used.

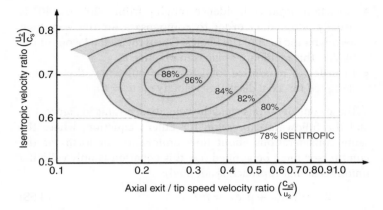

Figure 3.35 Attainable turbine efficiencies for different velocity ratios. *Source*: Rodgers and Geiser (1987) [18]. Reproduced with permission of the American Society of Mechanical Engineers.

maximum efficiencies are attained if the isentropic velocity ratio (u_2/c_s) is of the order of 0.68–0.71 [4]:

$$\frac{u_2}{c_s} = 0.68 - 0.71 \tag{3.70}$$

Furthermore, turbine efficiency can also be correlated with the exit velocity ratio (thus incorporating the effects of load and flow coefficients). In this case, a graphical representation of turbine efficiency similar to Smith's chart is obtained [18] (see Figure 3.35).

The foregoing considerations about optimum isentropic velocity ratio and exit velocity ratio, along with the recommendation for the incidence angle β_2, pose a restriction on the angle of the absolute flow leaving the stator vanes α_2, and accordingly the number of rotor vanes. Akin to the compressor impeller before, there is a criterion to select the optimum number of vanes, yielding a trade-off between flow guidance, friction losses and manufacturing costs.

The fundamental physics behind this problem was originally set forth by Jamieson who observed that too few blades could eventually lead to flow reversal due to the high tangential pressure gradient across the flow passage, in particular at very high values of α_2. Even if later experiments have observed that

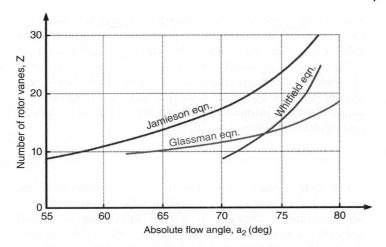

Figure 3.36 Minimum number of rotor blades for different absolute flow angles at rotor inlet.

the original number of blades reported by Jamieson was far too conservative, the approach has shown to be correct and thus it is still in use. This is observed in Figure 3.36, which compares Jamieson's results with those obtained by Glassman and Whitfield [6].

The number of blades actually results from the trade-off between:

- Flow guidance: flow reversal is avoided by increasing the number of blades.
- Friction losses: increasing the number of blades increases the total friction area, hence friction losses.
- Blockage losses: increasing the number of blades might result in excessive blockage at the rotor exit.

The latter effects can be tackled either by reducing the number of blades or by using half blades between the rotor inlet and halfway along the rotor channel. These are common design choices, in particular the former, as it does have a moderate impact on efficiency in particular if Jamieson's number of blades is used [4].

A final feature to be commented on is variable stator geometry. The backpressure imposed by the turbine on the upstream system (i.e. compressor and combustor) comes determined by the throat area of the stator, given that these devices are typically designed to operate above the critical pressure ratio (pressure ratio bringing about sonic conditions at the stator throat). The operation of a micro gas turbine at reduced loads is usually implemented by a lower rotational speed and mass flow rate, which inevitably brings about a lower pressure ratio even in the case of choked nozzles:

$$\frac{\dot{m}\sqrt{T_{01}}}{p_{01}} = \text{constant} \tag{3.71}$$

This lower pressure ratio has the following effects on the engine: lower cycle efficiency and specific output, and lower turbomachinery efficiency. It is thus interesting to achieve constant pressure ratio operation, which is enabled by reducing the throat area of the turbine stator at reduced flow conditions. Some concepts have been put forward in this regard, amongst which pivoting stator vanes is the most commonly used, in particular by the turbocharger industry. Unfortunately, the gap between the vanes and the side walls of the stator brings about clearance losses that reduce turbine efficiency. To avoid this, another solution is the use of movable sidewalls which reduce the axial height of the stator, keeping the blade stagger in its original position. Both solutions manage to keep β_2 at its optimum value, thus ensuring the highest turbine efficiency possible. Also, for a micro gas turbine, the engine benefits from a higher cycle efficiency.

The only negative feature of variable geometry turbines in micro gas turbines is compressor surge. Increasing pressure ratio at low flow coefficients reduces the surge margin and raises the risk of entering into unstable operation. These issues must be looked into when designing and matching the performance maps of compressor and turbine.

As was the case for compressors, the foregoing recommendations provide a preliminary design of the turbine, which can then be used as a starting point for further optimization. With this simple approach, a good estimate of turbine efficiency can be obtained by merely applying the non-dimensional coefficients introduced at the beginning of this section. A good overview is provided by the series of studies by Baines and coworkers

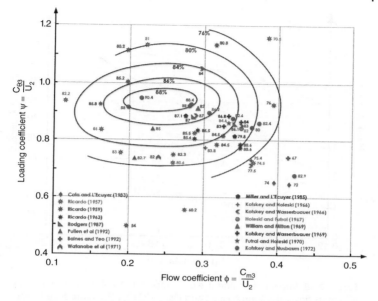

Figure 3.37 Attainable total-to-static efficiency depending on flow and loading coefficients. *Source*: Chen and Baines (1994) [21]. Reproduced with permission of Elsevier.

which evaluated the effects of stage loading on turbine efficiency [19, 20]. This was mentioned earlier with regard to Figure 3.35 and is now presented in Figure 3.37 [21].

The cited studies report that if the appropriate combination of stage loading and flow coefficient is selected, total to static efficiencies of the order of 0.8–0.85 are possible, with these values increasing to 0.90 for total-to-total efficiency. From this design, a numerical/experimental optimization can lead to a substantial improvement in the component performance [22], of the order of 2 to even 5 percentage points in efficiency.

3.2.5 Scaling of Radial Turbomachinery

Sometimes, in order to save time or reduce the computational effort needed for optimization, a previous compressor or turbine design is scaled up or down by merely applying the similarity parameters presented in a previous section. This is a common

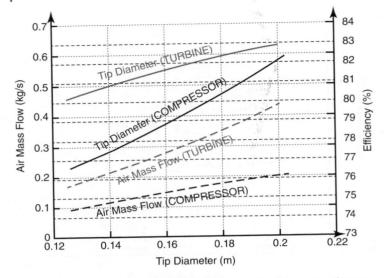

Figure 3.38 Impact of engine size on component total-to-static efficiency and mass flow rate. *Source*: Head and Visser (2012) [23]. Reproduced with permission of the American Society of Mechanical Engineers.

approach wherein the main differences between the existing references are found in the specific scaling laws. A detailed analysis is shown in [23] where the specifications of a Capstone C30 (30 kWe) engine and a 3 kWe scaled-down version are compared, Figure 3.38. It is observed that the impact of size on efficiency is not linear, due to the non-linearity introduced by the two dominant effects of size on component efficiency:

- Lower Reynolds number for similar Mach number at inlet and outlet.
- Larger clearances relative to impeller/wheel diameter, due to manufacturing limitations. A rough estimate of the impact on efficiency of clearances between impeller tip and casing in open impeller designs is provided by the following formula developed by Senoo and Ishida, integrating previous work in the topic:

$$\frac{\Delta\eta}{\eta} = 0.25\frac{t}{b_2} \tag{3.72}$$

where t is the average clearance between blade tip and casing and b_2 is the height of the blade at the impeller outlet [24].

These two effects have a negative influence on profile loss and tip clearance loss respectively, thus reducing the efficiency attainable by the compressor and turbine. More information about scaling methodologies and effects can be found in the cited handbooks.[12]

3.3 Recuperative Heat Exchanger

Section 3.1 put forward the need to employ a recuperative layout if high efficiencies are sought in micro gas turbines with inherent moderate pressure ratio and firing temperature. Such heat exchangers, preheating high-pressure combustion air and cooling down exhaust gases, must at the same time be efficient and bring about pressure losses that are as low as possible, given the very high sensitivity of these engines to pressure drops between compressor and expander. As discussed, this is due to the low useful to expansion work ratio that is characteristic of the Brayton cycle, aggravated by the usual working conditions of micro gas turbines. Finally, given that the recuperator contributes some 30% of the total cost of the engine, it is mandatory to reduce both material and manufacturing cost of this heat exchanger as much as possible.

Thus it is clear that there are several design requirements for recuperators, which are widely reviewed in [25]. Table 3.2 presents an excerpt of the cited reference [25], suggesting that the recuperator cannot be considered a secondary element relative to turbomachinery but, on the contrary, must be considered with equal importance.

There are several indirect-contact heat exchanger types that are eligible for the heat recovery duty in a micro gas turbine, and all of them share the aforesaid design constraints:

- Primary surface heat exchangers (PSHEs): the term 'primary surface' refers to the common heat transfer surface on both sides of the equipment; that is, there are no secondary surfaces

---

12 References [3, 4, 6, 8].

Table 3.2 Design requirements of micro gas turbine recuperators.

Area	Specific requirements
Major design criteria	Low heat exchanger cost Meet demanding micro turbine performance and economic goals High recuperator reliability
Performance	High recuperator effectiveness > 90% Low pressure drop < 5% Good part-load performance
Surface geometry	Primary surface geometry (no secondary surface inefficiency) High area/volume ratio (compactness) Superior thermal-hydraulic characteristics
Fabrication	Minimum number of matrix parts Continuous/automated fabrication process (mass production oriented) Welded sealing (eliminate need for furnace brazing)
Type of construction	Compact and lightweight matrix Integral manifolds/headers Matrix envelope flexibility (annular or platular)
Cost	No basic material wastage (zero scrap) Minimum (or zero) labour effort Standardization Case-specific materials selection Unit cost goal not to exceed 1.5 times material cost
Integrity	Resistant to thermal cycling Remain leak-tight for engine life Life goal of 50,000 h for micro turbine generator sets
Installation	Gas flow path compatibility with turbomachinery Compact and lightweight overall assembly Eliminate interconnecting ducts and thermal expansion devices

Table 3.2 (Continued)

Area	Specific requirements
Maintenance	Ease of recuperator removal/replacement
	Ease of leak detection testing
	Ease of weld repair
Performance growth capability	Adaptable to future higher temperature microturbine variants
	Materials selection flexibility (including ceramics)
	Adaptable to bi-metallic construction
	Retrofit capability with advanced heat exchanger concepts

Source: McDonald (2000) [25]. Reproduced with permission of Elsevier.

like pins or fins aimed at extending the heat transfer area on the side with lower film coefficient.

A primary surface recuperator can be manufactured by separating the hot and cold flows with a metallic sheet of a certain geometry. Also, in order to increase the heat transfer area per unit volume, inlet manifolds distribute both flows into a number of parallel streams which are typically arranged together into a sort of layered array, hence enabling multiple passages. Similarly, an outlet manifold collects all the outgoing flows in separate high-pressure and low-pressure streams. Annular or rectangular (plate), planar or wavy, there are multiple configurations enabling efficient heat exchange [26]. This is shown in Figure 3.39.

Primary surface recuperators have a number of advantageous features:

- Easy manufacturing process owing to the absence of internal welding or external brazing. The multiple primary surfaces are held together by simply welding along their edges, whilst the internal channelling (if any) is achieved by mere pressure contact between sheets [28]. This suffices to ensure leak-free operation.
- Virtually all the surfaces of the sheets separating the hot and cold flows effectively contribute to the heat exchange process, as opposed to other heat exchangers incorporating secondary surfaces [29].

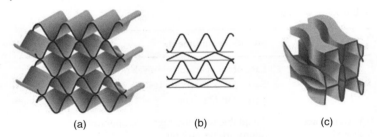

(a)   (b)   (c)

Figure 3.39 Different recuperator geometries. (a) Cross-corrugated, (b) wavy, (c) cross-wavy. *Source*: Utriainen and Sunden (2002) [27]. Reproduced with permission of the American Society of Mechanical Engineers.

- Owing to the possible stratification of temperatures, utilization of different material grades for the hot and cold ends is possible.
- Plate and fin heat exchangers (PFHEs): these heat exchangers fall into the compact heat exchanger class, namely heat exchangers with a very high specific heat transfer area (m^2/m^3). The cold and hot flows are separated by thin, flat plates with extended surfaces in between, and they can be arranged either in cross- or counter-flow layouts (Figure 3.40). These extended surfaces are of various types – flat, perforated, serrated/louvered, corrugated (simple or cross-corrugated), wavy – and they all have in common the aim to increase the heat transfer area and turbulence of the flow, in order to increase the heat flux across the fins (Figure 3.41). Also, from a global perspective, the heat exchangers can be manufactured in planar or annular geometries, depending on the system where they are incorporated. A difference with respect to primary surface heat exchangers is the need to braze the whole equipment together in order to ensure good thermal contact and leak-free operation. This is usually done in a vacuum furnace [30].

Two major shortcomings of these heat exchangers are:

- The likelihood of becoming fouled due to the very narrow passages which, along with the inherent difficulty of performing mechanical cleaning, makes them most suitable for clean working fluids.

- The lower efficiency of the secondary heat transfer area (fins), which increase the weight and cost. This higher cost adds up with that of the brazing process, hence increasing the specific cost of the technology.
- Tubular heat exchangers: as claimed by McDonald in [29], shell and tube heat exchangers might seem not appropriate for micro gas turbine applications where compactness seems to be a primary concern. Nevertheless, even if the bulky nature of tubular systems must be acknowledged, this technology exhibits other features that are also very interesting, as listed in Table 3.2. Amongst these, pressure containment capacity (not really an issue in conventional micro gas turbines given

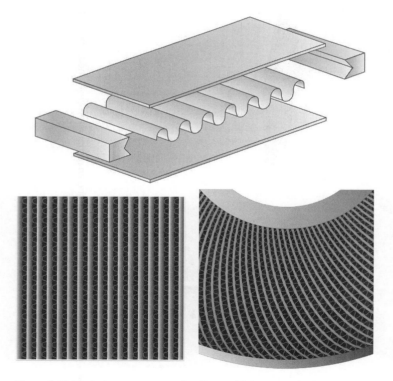

Figure 3.40 Typical arrangement of a plate and fin recuperator (top). Close-up of planar (bottom left) and annular (bottom right) layouts.

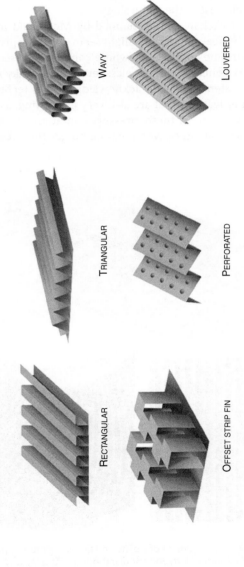

Figure 3.41 Different fin geometries in plate and fin recuperators (www.hedhme.com).

the low pressure ratio between 3:1 and 5:1) and resistance to thermal cycling are worth noting.

For these reasons, tubular technology as employed in recuperators for micro gas turbines makes use of very compact tubes with cylindrical or elliptical shapes and very small hydraulic diameters, fins also being an option if needed [26, 29].

A very interesting comparison is shown in Figure 3.42, extracted from [29]. The plot confirms that plate and fin

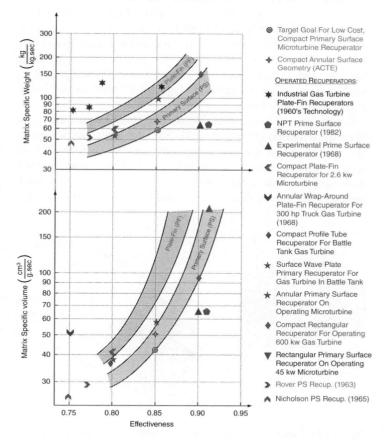

Figure 3.42 Specific area and cost for various recuperator technologies. *Source*: McDonald (2003) [29]. Reproduced with permission of Elsevier.

recuperators are bulkier and heavier than primary surface units of similar duty. Moreover, the relative dependence on effectiveness remains constant regardless of the latter parameter (duty), which means that the technology of choice should be the same for low-cost or high-performance applications. As a matter of fact, plate and fin systems seem to be unable to achieve the peak effectiveness attainable by primary surface recuperators.

Even if recuperator and regenerator are used interchangeably, they are formally different. In a recuperator, heat is transferred from the hot to the cold fluid directly through a surface, whereas in a regenerator an intermediate energy storage step takes place. Heat is transferred from the hot fluid to a movable heat storage device and then from this to the cold fluid flow (typically a solid ceramic matrix that is alternately put in contact with the hot and cold fluids by means of a rotary motion). Sample rotary heat exchangers are presented in Figure 3.43, showing both disk and drum type systems.

Regenerators have been widely used in small gas turbines – for instance vehicular gas turbines [31–33] – for their ability to manage waste heat in the exhaust gas stream. Nevertheless, these applications suffer from noteworthy flow leakages from the cold, high-pressure flow to the hot flow at much lower pressure. Such leakage can be as high as 15–20% and finds its way through the

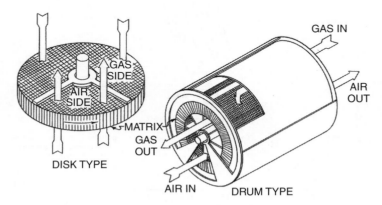

Figure 3.43 Sketch of a rotary regenerator. *Source*: Hammond and Evans (1956) [31]. Reproduced with permission of the American Society of Mechanical Engineers.

clearance gaps of the sliding seals, whose height comes about because of manufacturing limitations, wear and thermal expansion [34]. This constitutes the main limitation of rotary regenerators, along with the following design concerns:

- Characteristics of the regenerator matrix:
  - The thickness of the regenerator matrix has to be kept to a minimum in order to limit axial conduction losses and thermal stresses. This is particularly important for large temperature differences between cold and hot fluids (favoured by high turbine inlet temperature and low pressure ratio).
  - Porosity of the matrix is in the range from 40–70%. The former value provides higher durability and resistance to thermal fatigue, whereas the latter is suitable for lightweight engines.
  - Carryover flow is reduced by oversizing the seals so that they cover a larger portion of the matrix face area (as high as a fifth). Using an annular cross-sectional area with higher inner to outer diameter ratio also reduces these losses, even if this geometry might contribute to a larger engine footprint.
- Flow area split between hot and cold flows. The geometry of the matrix passages is common to both sides of the regenerator (given its rotary nature), but the density of these fluids is fairly different. It is therefore usual to have an uneven distribution of cross-sectional area which is typically larger on the hot side and smaller on the cold side, thus enabling similar pressure losses on both sides. This is typically defined by the conductance ratio C_r:

$$C_r = \frac{h \cdot A|_{hot}}{h \cdot A|_{cold}} \tag{3.73}$$

where h and A stand for heat transfer coefficient and cross-sectional area respectively.[13] The utilization of conductance ratio is a wise approach to area split, and has been used elsewhere for other applications and heat exchanger types [35, 36].

---

13 It must be noted that capacitance ratio is a design parameter for both recuperators and regenerators (i.e. it is not exclusive to the latter type of heat exchanger).

Wilson and Korakianitis provide recommended values for gas turbine recuperators [33]:
- Primary surface recuperators: $C_r = 1-4$.
- Rotary regenerators: $C_r \approx 1$.

These values are preliminary, though, inasmuch as adopting $C_r \approx 1$ would inevitably lead to uneven pressure drops across the hot and cold sides of the recuperator.

These and other recommendations are given in [33, 37] and [30] to produce a draft design of recuperators and regenerators which can then be iterated with help from numerical codes and laboratory testing. Moreover, such design must not only comply with the thermal duty, but also ensure mechanical integrity regarding creep life and oxidation/corrosion.[14]

In this latter regard, AISI 347 stainless steel is the most common material of choice in contemporary primary surface recuperators, as it is able to withstand temperatures close to 700°C (temperatures at expander exhaust are in the range 500–600°C due to the very low pressure ratio and despite the also moderate turbine inlet temperature of state-of-the-art engines) [38] and can be manufactured in the form of thin foils that can then be shaped as needed. For next generation higher exhaust temperature engines, the utilization of superalloys is also possible up to around 900°C, beyond which ceramic components are needed [39].

3.4 Bearings

The bearings in a micro gas turbine engine have the role to ensure that friction losses in the contact points between shaft and housing are as low as possible in order to increase mechanical efficiency. At the same time, these elements must be able to withstand the existing radial (typically very low) and axial (if any) loads.

There are different bearing technologies that can be used. The first type is magnetic bearings, which make use of electromagnetic forces to levitate a shaft rotating at high speed,

---

14 Even if there are other potential failures, these are acknowledged to be the most restrictive in usual applications.

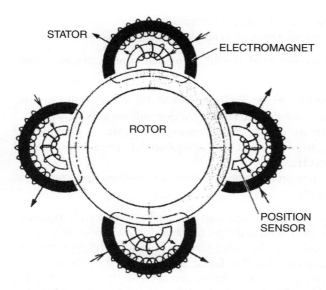

Figure 3.44 Active magnetic bearing schematic. *Source*: McDonald (1988) [40]. Reproduced with permission of the American Society of Mechanical Engineers.

thus ensuring minimum mechanical losses. These forces can be generated by permanent magnets (which are limited in temperature) or electromagnets, which consume a not-negligible fraction of the energy produced by the system but enable active control of the position of the shaft. In the latter case, a number of sensors track the relative position of the shaft and trigger corrective actions if necessary (hence the term active) by changing the inductance of the corresponding bearing stator. A schematic for an active magnetic bearing is shown in Figure 3.44.

The main advantages of active magnetic bearings are:

- Free contact, absence of lubrication and contaminating wear.
- Very narrow gap between rotor and housing ($\sim$0.1 mm).
- Low bearing losses, from 5–20% lower than ball and journal (fluid lubricated) bearings.
- The utilization of retainer bearings is possible. These act as emergency ball or journal bearings that come into service in case of malfunction/overload of the main magnetic bearings.

- Malfunctions are easy to diagnose.
- Low maintenance costs and long lifetime.

On the other hand, the following main limitations are identified:

- Specific load capacity limited to 30–60 N/cm^2, thus the maximum absolute load supported depends on bearing size.
- Maximum shaft speed limited to 6 kHz in commercial applications, even if the technology has been demonstrated at 300 kHz.
- High-temperature applications set further constraints on materials.

The second type of bearings that can be used in a micro gas turbine are the conformal fluid film bearings, wherein the rotating (shaft) and static (housing) surfaces are completely separated by a lubricant film, whether liquid or gas. When the pressure of the lubricant is provided by the motion of the shaft itself, the bearing is called a self-acting or hydrodynamic bearing. If, on the other hand, it is an external device pressurizing the lubricant, the bearing is of the hydrostatic type.

Again, there is no contact between the stationary and rotating surfaces when a hydrodynamic bearing is used, as a fluid film is permanently separating both surfaces. The thickness of this film (h_{film}) exhibits the following dependence on rotating speed (v) and load (N):

$$h_{film} \propto \sqrt{\frac{v}{N}} \tag{3.74}$$

It is easily deduced that the performance of the bearing depends on the viscosity of the fluid, both for the load-carrying capacity and the friction losses (fluid shearing). Amongst the various types of oil-lubricated bearings, Soares reports the floating sleeve valve as the most common [41]. These use high-grade oil lubricant which enables extremely long life, as the risk of oil contamination is minimal. Also gas bearings are becoming increasingly popular, for they do not need a separate lube oil system (oil storage, filtering, pumping and piping), and nor do they have sealing requirements (i.e. a dedicated system to retain the lubricant within the bearing containment) [42].

Within the general category of sleeve bearings, tilting pad is the most common bearing type in contemporary turbomachinery [43]. A number of pads are inserted in the gap between the rotating shaft and the journal support, with a tilting degree of freedom that enables them always to work in the best position [44].

One of the main concerns during the design phase of an oil bearing is the choice of adopting hydrostatic or hydrodynamic operation. The former ensures contact-free operation during the startup and shutdown phases, but at the expense of installing a lubricant pressurizing system which means higher costs and potentially lower reliability. Hydrodynamic bearings do not need such a system, but this is at the cost of incurring additional wear on the contact surfaces between shaft and support when the system rotates at very low speed (typically startup and shutdown manoeuvres).

Regarding the ideal lubricant, the utilization of oil in micro gas turbines was typical in the past, despite higher mechanical losses (in particular during startup owing to the lower lube oil temperature) and maintenance costs, and requiring additional auxiliary equipment (oil pump, cooler, filter, sump tank). For these reasons, and in spite of oil lubrication being still used by some manufacturers (see section below), gas bearings have recently become the dominant technology in these applications, either tilting pad or foil bearings.

Common gas foil bearings are depicted in Figure 3.45, showing Gen I leaf bearings and more modern bump-type units. Taking the bottom left picture as reference, the bump foil is subjected to a certain load when the shaft is at rest, the top foil thus exerting pressure on the shaft. Then, as the engine spins, the hydrodynamic effect of air in the gap between top foil and shaft lifts the latter, which becomes suspended on a gas film. This operating regime drastically reduces friction losses and enables thermal expansion of the shaft, which is accommodated by pushing the foil radially outwards for a constant gas film thickness. Actually, this latter feature is extremely important in micro gas turbines, given that the shaft experiences a significant temperature change during startup and load changes (in particular the former manoeuvre) due to the compactness of

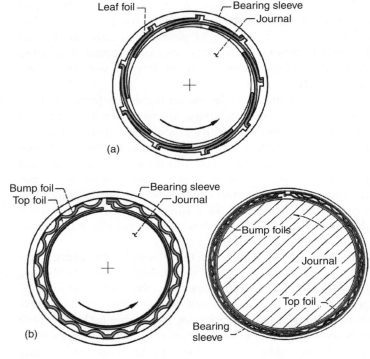

Figure 3.45 a) Standard Gen I leaf bearing. b) Gen I (left) and Gen III (right) bump bearings [45].[15]

the system and large temperature gradients in short distances (typically tens of centimetres).

So far the discussion has focused on journal bearings, meaning elements that are subjected to radial loads mainly due to the effect of gravity. The concept of foil bearings can nevertheless

---

15 Gen I and III refer to the particular foil design and, accordingly, the load-carrying capacity at a given speed. Gen I bearings use a simple foil whose properties are constant in all directions and whose load carrying capacity is rather small, mainly due to end leakage of the working gas. Contemporary Gen III bearings makes use of several foils assembled in a more complex design yielding variable properties in each direction and enabling higher load-carrying capacity at a given shaft speed.

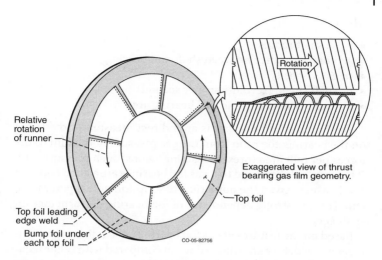

Figure 3.46 Standard geometry of a Gen III foil thrust bearing [46].

be extended to elements subjected to axial loads, namely thrust bearings, which can adopt the same conceptual scheme. Such a foil thrust bearing is shown in Figure 3.46 for a last generation unit.

Gas foil bearings are currently considered the most appropriate technology for micro gas turbines. These are actually preferred to roller-element bearings (due to the very high shaft speeds) or oil bearings (due to the lower maintenance requirements and simpler balance of plant). The latter technology is nonetheless used sometimes, as already mentioned, in particular in those applications where the shaft arrangement is adapted from turbochargers which, in the main, make use of oil bearings taking advantage of the reciprocating engine's lube oil system [47]. A simple means to identify whether gas foil bearings are suitable for an application is the Sommerfeld number:

$$S = \left(\frac{r_{shaft}}{e} \right)^2 \frac{\mu \omega}{P} \approx \frac{\text{hydrodynamic pressure}}{\text{specific load per unit area}} \qquad (3.75)$$

where:

- r_{shaft}: shaft radius
- e: radial clearance between shaft and bearing
- μ: gas viscosity
- ω: shaft speed (revolutions per second)
- P: load per unit of projected bearing area.

The Sommerfeld number is an indirect measure of the stress on the bearing: the ratio from the gas pressure in the film to the forces per unit area (pressure) acting against such pressure [48]. It is commonly assumed that gas foil bearings are a good choice if $S > 6$, which, given the numerical description of this parameter, confirms the strong dependence of foil bearing performance on shaft speed.

Based on the features of gas foil bearings, the following advantages are widely acknowledged when compared with other technologies like roller-elements, magnetic or oil-lubricated bearings [42, 48]:

- Higher reliability due to the presence of foils that minimize the potential wear due to contact between solid surfaces.
- No scheduled maintenance due to the absence of a lubricant system. This is particularly relevant when compared with roller bearings in high-speed applications.
- High-speed capability, boosted by the hydrodynamic effect of gas at high speed.
- Low- and high-temperature capabilities, enabling large thermal expansions of the main shaft.
- High stability owing to the damping effect of the foil pads, hence preventing whirl events.

Amongst the main disadvantages, the lack of standardization and commercial availability may be noted. This means that a new, case-specific design is mandatory for each application [49].

The aforementioned features explain why foil bearings are now the technology of choice in many micro gas turbine applications, for instance Capstone, even if other engines like Ansaldo's Turbec T100 or Ingersoll-Rand's MT250 Series still rely on oil-lubricated bearings owing to the proven reliability of the technology.

3.5 Conclusions: Commercial Status and Areas of Research

The micro turbine industry is experiencing a new renaissance after several decades of idling, mainly thanks to the strong support received by distributed generation initiatives. Unfortunately, there is not a consolidated industry to readily deploy the technology. A number of companies entered the market in the last decade but were forced to quit it after realizing there was not a large enough market volume to share and, most importantly, there was no standardization on the customer's side. Some noteworthy examples are Elliot Energy Systems (later Calnetix) and Bowman Power.

As of today, the number of companies offering commercial micro gas turbines products is small. Capstone Inc. (www.capstoneturbine.com) dominates the market and has arguably the most reliable and competitive product range, the C series offering engines with 30, 65 and 200 kWe electric outputs. The latter can also be packed together as a single set for a total of 600, 800 or 1000 kWe system output. Capstone's systems adapt to a variety of applications from stand-alone power generation to combined heat and power, and can operate on a number of standard (methane, diesel, kerosene) and low calorific value (biogas, landfill gas, flare gas) fuels.

Ingersoll-Rand offers the MT250 unit with similar flexibility and reliability as Capstone's products (www.ingersollrandproducts.com). The product range is narrower and limited to a 250 kWe engine but, from an end-user perspective, it has the same capability to adapt to different applications (combined heat and power or power-only) and fuels.

Turbec emerged as the European counterpart of Capstone. Currently owned by Ansaldo (www.ansaldoenergia.it), it commercializes a 100 kWe unit with reliability proven by a large number of units in operation in a variety of applications. The standard product line offers units for biogas, natural gas and external heating, which is an innovative feature of this system (given that it is in the product line and not on demand). The Eco+Energy Compact Power System manufactured by Dürr in Germany is very similar in concept and specifications to the T100 engine by Ansaldo. It delivers 100 kWe in combined heat and power applications with 30% electrical efficiency (www.durr-cleantechnology.com). Moreover, akin to the Italian engine, the unit delivered by Dürr can be operated in external heating mode with solid fuels.

Just recently, Micro Turbo Technology MTT in the Netherlands (www.mtt-eu-com) has launched a combined heat and power unit based on a 3 kWe micro gas turbine rotating at very high speed. This system is already certified and there are plans for a scale-up to the 15–20 kWe range. In a similar output range, the UK-based company Bladon Jets commercializes an aeroderivative generator set rated at 12 kWe. This system is reportedly aimed at providing power for telecommunication stations.

The main specifications for these engines are summarized in Table 3.3, where the following similarities and differences are worth noting:

- All the engines make use of a recuperative layout to save fuel by recovering the sensible heat in the exhaust gas stream.
- Capstone, Bladon Jets and MTT make use of gas bearings, whilst Ingersoll-Rand and Ansaldo employ lube oil bearings. This is an added operating cost to the system which, in the case of the T100 engine, amounts to 3 litres of oil for every 6000 operating hours. There is no information reported by MTT in this respect.
- The values reported for turbine exhaust temperatures are very similar. This suggests that all the engines operate with very similar pressure ratios and turbine inlet temperatures (i.e. working cycle), even though these values are reported by Ansaldo and (partially) Dürr only.
- The scale effects on efficiency are evident in the series of engines by Capstone, where a large fraction of the gain in performance is attributable to this influence.

The foregoing considerations about the market status indicate that several areas of technology require further research. This research can be divided in the following categories:

- *Turbomachinery*. The main challenges concern manufacturability of the very small compressors and turbines. Additive manufacturing is seen as a very attractive option, even if the resistance to thermal stress in the turbine is still uncertain. Also in this area, the utilization of variable geometry components requires further investigation.

- *Recuperator.* Given the very large share of this component in the total cost of the engine, new manufacturing techniques have to be developed enabling a large cost reduction that could significantly impact the manufacturing costs.
- *Materials.* It is very likely that ceramics will have to be used in next-generation micro gas turbines to exploit the benefits of higher efficiency and specific output without incurring the costs associated with more complex cycle layouts. If these materials were used, new recuperator materials and manufacturing techniques would have to be developed.
- *Bearings.* The lack of standardization of gas bearings has already been highlighted as one major drawback of this technology, which inevitably brings about higher engineering costs. More research is needed in this area.
- *Power electronics.* High-speed alternators are currently very expensive. Further research is needed to drive their cost down.

All these topics have been highlighted throughout the chapter and numerous references have been given. Nevertheless, these

Table 3.3 Commercial micro gas turbines.

Manufacturer	Model	$\dot{W}_e$ [kWe]	η_{net} [%]	*TET* [°C]	Features
Capstone	C30	30	26	275	Gas bearings
	C65	65	29	309	
	C200	200	33	280	
Ingersoll-Rand	MT250	250	30	242	Lube oil bearings
Ansaldo	T100	100	30	270	Lube oil bearings
					Turbine inlet 950°C
					Pressure ratio 4.5
Durr	CPS	100	30	270	Pressure ratio 4.5
Bladon Jets	MTG12	12	26.5	–	Gas bearings
MTT	EnerTwin	3	15	–	Rated shaft speed 240 krpm

are just the well-identified areas where there is consensus in the need to develop the technology further. Occasionally, though, disruptive ideas appear, like the rotating combustor under development at MTT [50]. These are of course welcomed as they introduce step-changes that bridge the gap to full commercialization of cost-competitive products. Along the same lines, a survey of new applications like, for instance, automotive range extenders, helps the industry to gain critical mass and thus take advantage of economies of scale along the entire product supply chain.

3.6 Questions and Exercises

1 The following assumptions are made in order to produce a draft design of a new micro gas turbine: turbine inlet temperature (900°C), pressure ratio (3.5), pressure drop across inlet filter (10 mbar), pressure drop across the combustor (2.5%), isentropic efficiency of compressor (0.83) and expander (0.88). The engine operates on a simple Brayton cycle arranged in a single shaft (compressor-expander generator) whose mechanical efficiency is estimated at 95%. Provide a first estimate of the expected thermal efficiency (in ISO conditions) of the engine. If the engine is rated at 100 kW (shaft output), estimate the approximate air mass flow rate at compressor inlet. Heat losses can, at this stage, be considered negligible, whereas air and combustion gases are assumed to behave ideally with the following properties: $c_{p,air} = 1.005$ kJ/kg·K, $\gamma_{air} = 1.4$, $c_{p,gas} = 1.148$ kJ/kg·K, $\gamma_{gas} = 4/3$.

2 Given the low efficiency of the engine, the design team considers introducing a recuperator with an estimated efficiency of 85%. If the expected pressure drop on each side of this equipment is 1.5%, what would be the efficiency of the recuperated engine, all other parameters remaining equal? Would the mass flow rate at the compressor inlet change?

3 Analyse the temperature of the working fluid at each station of the engine. Which material would you select to manufacture the recuperator?

4 In the process of designing the compressor, a first tentative design considers a single stage compressor using a backswept impeller and vaneless diffuser. The following assumptions are made: inlet velocity $c_1 = 85$ m/s (no swirl), relative tip Mach number at the inlet $M'_{1t} = 0.85$, absolute Mach number at impeller outlet $M_2 = 0.95$, diffuser efficiency $\eta_d = 0.78$, aspect ratio at impeller outlet $b_2/r_2 = 0.1$. With this information, complete the draft design of the compressor impeller (inlet and outlet radii of impeller and diffuser, blade height at impeller outlet, stage reaction, shaft speed, outlet sweep angle).

5 Starting from scratch, would you be able to produce a draft design of the expander driving the compressor and electric generator? Be explicit about packaging-related considerations (i.e. common sizes of turbomachinery envelope).

6 In the light of the resulting shaft speed, what bearing technology would be most suitable? What would be the main disadvantages of using lube-oil bearings?

References

1 Haywood, R.W. (1991) *Analysis of Engineering Cycles* (4th edn). Pergamon Press, Oxford.
2 Saravanamuttoo, H.I.H., *et al.* (2009) *Gas Turbine Theory* (6th edn). Pearson Education Limited, Harlow.
3 Balje, O.E. (1981) *Turbomachinery: A Guide to Design, Selection and Theory.* John Wiley and Sons, Ltd., Chichester.
4 Dixon, S.L. (1978) *Fluid Mechanics and Thermodynamics of Turbomachinery* (3rd edn). Pergamon Press, Oxford.

5 Krain, H. (2005) Review of centrifugal compressor's application and development. *Journal of Turbomachinery*, 127, 25–34.

6 Whitfield, A. and Baines, N.C. (1990) *Design of Radial Turbomachines*. Longman Scientific and Technical, Harlow.

7 Brown, W.B. and Bradshaw, G.R. (1947) *Method of designing vaneless diffusers and experimental investigation of certain undertermined parameters TN-1426*. Flight Propulsion Research Laboratory, NACA. NACA, Cleveland.

8 Japikse, D. (1996) *Centrifugal Compressor Design and Performance*. Concepts ETI, Inc., Wilder.

9 Sanchez Lencero, T., Muñoz Blanco, A. and Jimenez-Espadafor, F.J. (2004) *Turbomáquinas térmicas*. Síntesis, Madrid.

10 White, R.C. and Kurz, R. (2004) Surge avoidance in gas compression systems. *Journal of Turbomachinery*, 126, 501–506.

11 Sentz, R.H. (1980) The analysis of surge. 9th Turbomachinery Symposium, Texas A&M University, Austin TX, 57–62.

12 McMillan, G.K. (2010) *Centrifugal and Axial Compressor Control* (1st edn). Momentum Press, New York.

13 Brun, K. and Nored, M.G. (2008) *Application Guideline for Centrifugal Compressor Surge Control Systems*. Gas Machinery Research Council. South West Research Council, Dallas, TX.

14 Simon, H., Wallmann, T. and Monk, T. (1986) Improvements in performance characteristics of single-stage and multistage centrifugal compressors by simultaneous adjustments of inlet guide vanes and diffuser vanes. 31st International Gas Turbine Conference and Exhibit, ASME-IGTI, Dusseldorf.

15 Scotti Del Greco, A. and Tapinassi, L. (2013) On the combined effect on operating range of adjustable inlet guide vanes and variable speed in process multistage centrifugal compressors. ASME Turbo Expo, ASME-IGTI, San Antonio.

16 De Paepe, W., *et al.* (2014) T100 micro gas turbine converted to full humid air operation: test rig evaluation. ASME-IGTI, ASME Turbo Expo: Turbine Technical Conference and Exposition, Dusseldorf.

17 Pezzini, P., Tucker, D. and Traverso, A. (2013) Avoiding Compressor Surge During Emergency Shut-Down Hybrid

Turbine Systems. ASME-IGTI, Proceedings of Turbo Expo, San Antonio.

18 Rodgers, C. and Geiser, R. (1987) Performance of a high-efficiency radial/axial turbine. *Journal of Turbomachinery*, 109, 151–154.

19 Baines, N.C. (1996) Flow development in radial turbine rotors. Proceedings of the International Gas Turbine and Aeroengine Congress & Exhibition, Birmingham.

20 Chen, H., et al. (1992) The effects of blade loading in radial and mixed flow turbines. Proceedings of the International Gas Turbine and Aeroengine Congress and Exposition, Cologne.

21 Chen, H. and Baines, N.C. (1994) The aerodynamic loading of radial and mixed-flow turbines. *International Journal of Mechanical Science*, 36, 63–79.

22 Barsi, D., et al. (2015) Radial inflow turbine design through multi-diciplinary optimisation technique. Proceedings of the Turbo Expo: Turbine Technical Conference and Exposition, Montreal.

23 Head, A.J. and Visser, W.P.J. (2012) Scaling 3-36 kW microturbines. ASME-IGTI, Proceedings of Turbo Expo, Copenhagen.

24 Senoo, Y. and Ishida, M. (1987) Deterioration of compressor performance due to tip clearance of centrifugal impellers. *Journal of Turbomachinery*, 109, 55–61.

25 McDonald, C.F. (2000) Low-cost compact primary surface recuperator concept for microturbines. *Applied Thermal Engineering*, 20, 471–497.

26 Shah, R.K. (2005) Compact heat exchangers for microturbines. 5th International Conference on Enhanced Compact and Ultra-Compact Heat Exchangers: Science, Engineering and Technology. Engineering Conferences International, Hoboken, NJ, 247–257.

27 Utriainen, E. and Sunden, B. (2002) Evaluation of the cross corrugated and some other candidate heat transfer surfaces for microturbine recuperators. *Journal of Engineering for Gas Turbines and Power*, 124, 550–560.

28 Traverso, A. and Massardo, A.F. (2005) Optimal design of compact recuperators for microturbine applications. *Applied Thermal Engineering*, 25, 2054–2071.

29 McDonald, C.F. (2003) Recuperator considerations for future higher efficiency microturbines. *Applied Thermal Engineering*, 23, 1463–1487.

30 Kakaç, S. and Liu, H. (2002) *Heat Exchangers: Selection, Rating and Thermal Design* (1st edn). CRC Press, Boca Raton.

31 Hammond, W.E. and Evans, T.C. (1956) Gas-turbine rotary regenerator – Design and development of prototype unit for 3000 hp plant. Proceedings of the ASME Gas Turbine Power Conference, Washington.

32 McDonald, C.F. (1997) Ceramic heat exchangers – The key to high efficiency in very small gas turbines. Proceedings of the International Gas Turbine & Aeroengine Congress & Exhibition, Orlando.

33 Wilson, D.G. and Korakianitis, T. (1998) *The Design of High-efficiency Turbomachinery and Gas Turbines* (2nd edn). Prentice-Hall, Upper Saddle River.

34 Kluka, J.A. and Wilson, D.G. (1998) Low-leakage modular regenerators for gas-turbine engines. *Journal of Engineering for Gas Turbines and Power*, 120, 358–362.

35 Ameel, T.A. (2000) Parallel-flow heat exchangers with ambient thermal interaction. *Heat Transfer Engineering*, 21, 18–25.

36 Hoopes, K., Sanchez, D. and Crespi, F. (2016) Modelling off-design performance of sCO_2 heat exchangers. Proceedings of the 5th International Symposium on Supercritical CO_2 Power Cycles, San Antonio, TX.

37 Kays, W.M. and London, A.L. (1984) *Compact Heat Exchangers* (3rd edn). McGraw-Hill, New York.

38 Omatete, O.O., *et al.* (2000) *Assessment of Recuperator Materials for Microturbines*. Department of Energy, ORNL/TM-2000/304. Oak Ridge National Laboratory, Oak Ridge.

39 McDonald, C.F. (1996) Heat recovery exchanger technology for very small gas turbines. *International Journal of Turbo and Jet Engines*, 13, 239–261.

40 McDonald, C.F. (1988) Active magnetic bearings for gas turbomachinery in closed-cycle power plant systems. ASME-IGTI, Gas Turbine and Aeroengine Congress, Amsterdam.

41 Soares, C. (2007) *Microturbines* (1st edn). Academic Press, Burlington.

42 Agrawal, G.L. (1997) Foil air/gas bearing technology – an overview. ASME-IGTI, International Gas Turbine and Aeroengine Congress and Exhibition, Orlando.

43 Boyce, M.P. (2003) *Centrifugal Compressors* (1st edn). PennWell, Tulsa.

44 Salamone, D.J. (ed.) (1984) *Journal bearing design types and their application to turbomachinery*. 13th Turbomachinery Symposium. Texas A&M University Press, College Station.

45 DellaCorte, C., Zaldana, A.R. and Radil, K.C. (2004) A systems approach to the solid lubrication of foil air bearings for oil-free turbomachinery. *Journal of Tribology*, 126, 207.

46 Dykas, B., *et al.* (2008) Design, fabrication, and performance of foil gas thrust bearings for microturbomachinery applications. *Journal of Engineering for Gas Turbines and Power*, 131, 012301-8.

47 Visser, W.P.J., Shakariyants, S.A. and Oostveen, M. (2011) Development of a 3 kW microturbine for CHP applications. *Journal of Engineering for Gas Turbines and Power*, 133, 042301-8.

48 DellaCorte, C., *et al.* (2006) *A Preliminary Foil Gas Bearing Performance Map*. Glenn Research Center, NASA, Technical Memorandum TM2006-214343. Society of Tribologists and Lubrication Engineers, Cleveland.

49 Barber-Nichols Inc. (2013) *Guidelines for Determining Foil Bearing Applicability*. Technical note. Barber-Nichols Inc., Arvada. www.barber-nichols.com

50 Kornilov, V.N., Shakariyants, S. and de Goey, L.P.H. (2012) Novel burner concept for premixed surface-stabilized combustion. Proceedings of the Turbo Expo, Copenhagen.

4

SOFC/mGT Coupling

The primary outstanding issues of hybrid systems are related not only to SOFC aspects (e.g. cost reduction, reliability increase, material issues, etc.), but also to critical features of the coupling between the stack and the micro gas turbine. For this reason, this chapter starts with SOFC hybrid system concept development, covers various aspects of ongoing research focusing on SOFC/mGT systems, and concludes by briefly describing the currently existing hybrid plant prototypes developed (or under development) by commercial entities.

4.1 Basic Aspects of SOFC Hybridization

During the 1990s, several authors presented efficiency improvements and perceived operational improvements related to SOFC hybrid systems [1–4], noting the beneficial operational

Hybrid Systems Based on Solid Oxide Fuel Cells: Modelling and Design, First Edition.
Mario L. Ferrari, Usman M. Damo, Ali Turan, and David Sánchez.
© 2017 John Wiley & Sons Ltd. Published 2017 by John Wiley & Sons Ltd.

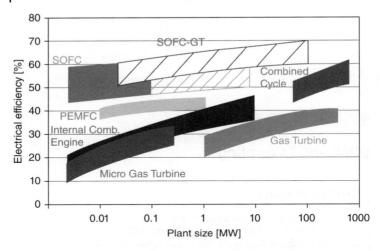

Figure 4.1 Electrical efficiency of some fossil-fuel based plants.

experience regarding combined cycles due to the utilization of the high temperature exhaust flow produced by the SOFCs. Initial theoretical calculations carried out on these new plants showed an important performance trend: increasing efficiency and reducing CO_2 emissions. These beneficial attributes can be so great that hybrid systems can be qualified as the 'most efficient power plants' based on fossil fuels. SOFC hybrid systems not only display positive large-scale performance, but also allow for smaller-size power plants which can be used for distributed power generation. These important features are shown in Figure 4.1, where the electrical efficiency of some fossil-fuel based traditional plants is compared with the same performance parameters for SOFC hybrid systems (just the plants based on the coupling of SOFCs with gas turbines are included, as this is the most promising technology, as discussed below). While the forms in solid fill refer to existing technology, those filled with a hached pattern are included for systems not yet ready for commercialization (e.g. SOFC-GT power plants).

4.2   SOFC Coupling with Traditional Power Plants

Several traditional power plants were taken into account for the coupling with SOFC stacks to maximize the overall plant efficiency [3]. Initially, these calculations were developed considering primarily the main components at design conditions for the development of large scale plants (exploiting the positive benefits in terms of efficiency for large scale plants). However, these initial results showed too high efficiency values, as demonstrated by further calculations based on more realistic operating conditions (e.g. including the auxiliaries and considering a fuel cell sized on a cost consideration basis), or using more feasible layouts for component constraints (see paragraph 4.3). So, in the following paragraphs attention is focused on possible basic plant layouts, starting from initial theoretical results [3].

4.2.1   Coupling with Steam Power Plants

Starting from the basic concepts developed for combined cycles, an initial layout considered to improve SOFC efficiency is based on the coupling of a stack with a steam power plant [3, 5]. As in standard combined cycles, the steam power system is the bottoming cycle (Figure 4.2) able to exploit part of the thermal energy of the SOFC exhaust flow. So, the fuel cell works at atmospheric conditions (with a cathodic recirculation essential for air pre-heating) and delivers the exhaust flow to a heat recovery steam generator (HRSG) downstream of the off-gas burner (OGB), which is an essential component to burn the leftover fuel flowing out from the anode.

Preliminary calculations have shown the beneficial influences related to the coupling of these systems, obtaining an electrical efficiency close to 72% (with 79% of the power produced by the stack) [3]. However, such conceived outlays of plants require a very large size of stack (which is currently not available on the market) for coupling with the large steam power plant (the bottoming cycle). Moreover, the calculated efficiency does not represent a realistic net performance indicator, since it is not

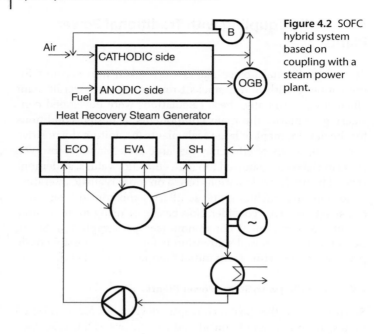

Figure 4.2 SOFC hybrid system based on coupling with a steam power plant.

completely inclusive of all the consumed power, such as the fuel processing issues (energy consumed in the reforming reactions) or the other auxiliary components. A further important coupling issue to be considered for these systems relates to the various local temperatures: the off-gas burner (OGB) outlet temperature could be too high (e.g. higher than 650°C) for the heat recovery steam generator (HRSG) inlet, in cases where a high-temperature SOFC is considered [3]. A possible solution could be devised by using this flow to pre-heat both the fresh air and the fuel flow. However, this proposed solution has to be assessed with regard to the power consumed for reforming reactions. A significant reduction in efficiency is expected compared with the simplified theoretical calculation.

4.2.2 Coupling with Gas Turbines

Since coupling with a simple cycle system is not able to maximize the operational efficiency, preliminary theoretical calculations were carried out on an SOFC system coupled with

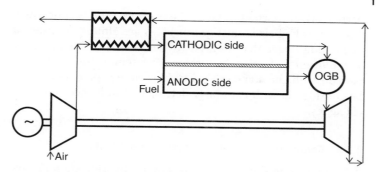

Figure 4.3 SOFC hybrid system based on coupling with a recuperated gas turbine.

a recuperated gas turbine [3]. As shown in Figure 4.3, the fuel cell was connected to the turbine by placing it where there would typically be a combustion chamber (on the air side). So, the compressed air is fed to the SOFC stack (cathode side) downstream of the counterflow recuperator (air side). Then, the anodic exhaust gas (including an incompletely oxidized fuel flow) is mixed with the cathodic outlet flow in the off-gas burner located upstream of the expander. Finally, this exhaust gas is used on the hot side of the recuperator [6].

The initial theoretical calculations for this particular design of plant layout produced an electrical efficiency close to 82%, with about 69% of the electrical power produced by the fuel cell [3]. However, these results are now considered too optimistic as they are not completely based on realistic operational and design constraints, such as fuel processing issues (e.g. steam to be injected or recirculated for reforming reactions) or other component constraints (e.g. a turbine outlet temperature higher than 650°C implies the application of expensive materials for the recuperators). Moreover, a more realistic assessment needs to take into account the auxiliary power absorbed with much more care. A further important consideration that affects hybrid system performance relates to specific turbine parameters. Indeed, there are no commercially available gas turbines specifically developed for SOFC applications (i.e. designed with working cycles that optimize the efficiency of the hybrid system rather than the gas turbine itself). This is particularly important as far as the pressure ratio is concerned. Also, regarding size, most gas turbines are almost an order of magnitude larger than desired (MWe vs. kWe). Furthermore, a significant reduction in turbine inlet temperature (in comparison with the standard gas turbine design practice) has to be considered, with extremely detrimental influences on cycle efficiency.

4.2.3 Coupling with Combined Cycle-based Plants

A final coupling solution considers SOFC hybridization involving a combined cycle system [3]. Initially, this could present the best option in terms of efficiency, as it is based on a coupling practice involving the most promising fuel cell technology with the highest efficiency conventional power generation technology. Moreover, the global system would be composed of three integrated power plants able to operate at high, intermediate and low temperatures with a beneficial effect regarding energy loss minimization.

As shown in Figure 4.4, the fuel cell system (including the OGB) is located between the compressor and the turbine. However, since in this layout no recuperator is included, the air stream has to be pre-heated using part of the heat produced by the fuel cell in an internal heat exchanger (HEx). Then, the OGB exhaust flow is conveyed to the gas turbine with a similar approach as shown in Figure 4.3. Finally, the exhaust stream from the expander is in the 600–650°C temperature range, and thus has a very high enthalpy that can be used further in the HRSG and associated steam plant.

Initial theoretical calculations revealed a very high efficiency value (close to 76%, with 75% of the output produced by the fuel cell [3]), showing the positive effects related to the combined cycle in comparison with coupling the SOFC with a simple steam power plant. However, in this case some constraints and losses (e.g. the power consumed by the auxiliaries) were not realistically taken into account. Moreover, this final hybrid solution is particularly affected by substantial fundamental disadvantages involving both traditional plants: no gas turbines have been specifically designed for SOFC applications and there is a very large stack size requirement.

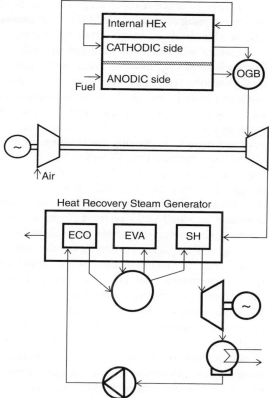

Figure 4.4 SOFC hybrid system based on coupling with a combined cycle system.

4.3 Beneficial Attributes Related to SOFC/mGT Coupling

Even though the previous paragraphs describe the very significant performance increases achievable by using SOFCs operating in connection with large-scale traditional power plants, several design and operational issues regarding fuel cell technology development are limiting the SOFC size to hundreds of kilowatts. Driven primarily by cost, reliability and control issues, SOFC manufacturers are focusing part of their research and development (R&D) activities on the optimization of small-scale reactors, along with developing designs of fuel cells able to produce realistic megawatt size plants [3]. Even if a simple coupling of small-scale SOFCs can produce a large-scale reactor (hundreds of megawatts), no commercial R&D effort is currently pursued by companies in such undertakings [7]. For such reasons, most current generation hybrid system configurations related to large-scale power generation systems (such as plants based on the coupling between an SOFC stack with a hundred-megawatt gas turbine, a steam power plant or a combined cycle) are not under development or planned to be designed prior to improving the aforementioned SOFC issues.

Micro gas turbine (mGT) technology has been adopted as a promising intermediate solution for initial system prototype designs involving SOFC hybrid systems [3, 7, 8]. Specifically, they show promise in this effort as microturbines are characterized as small-scale systems (from a few to hundreds of kilowatts) operating with mass flow values in the same range as existing SOFC stacks. Furthermore, the coupling engineering issues for these machines cover SOFCs operating at both atmospheric or pressurized (4–7 bar range) modes. Additionally the turbine inlet temperature (TIT) values for such systems are close to stack discharge conditions (which is important for radial commercial microturbines with no blade cooling, as the SOFC system outlet temperature values are close to design turbine inlet conditions). An important consideration involves the placement of the off-gas burner (OGB). In the case of low-temperature SOFCs, this component has to be installed between the stack and the turbine (to have the correct TIT value), but for high-temperature stacks the OGB could be located in a recirculation line to pre-heat the cathodic inlet flow. Another important advantage relates to the available cathode inlet temperature. Specifically, since mGTs are mainly based on a recuperated cycle, the airflow is pre-heated (as necessary for a coupling with an SOFC) using the exhaust thermal energy [9]. While this solution is adequate for tubular SOFCs (equipped with an internal air heating pipe), in the case of a high-temperature planar SOFC, a further pre-heating step could be required (usually carried out with a cathodic recirculation system [10]). Finally, the SOFC/mGT coupling can further benefit from a more straightforward electrical system coupling. Specifically, since microturbines operate at very high rotational

Table 4.1 SOFC/mGT coupling advantages and disadvantages.

Advantages
Size and airflow values consistent with SOFC stack size
Pressure value consistent with pressurized existing SOFC stacks
Turbine inlet temperature values close to stack discharge conditions
Available air temperature values close to acceptable SOFC cathode inlet temperature range, if the microturbine is based on the recuperated cycle
Possible electrical integration at continuous current level
Disadvantages
Commercial microturbines not specifically designed for SOFC issues
Significant influence of ambient temperature value
Plant exhaust flow temperature cannot be decreased to values lower than 200–250°C (however, this disadvantage could be a positive aspect in case of cogeneration)
Controllability of dynamic phenomena
Integration of components with different dynamic timescales

speeds, a current rectifier is usually installed upstream of the inverter necessary to connect the machine with the electrical grid. A direct current line between the two power electronic systems allows (if the stack is properly designed) a simpler connection with the SOFC electrical system that operates in the direct current mode [3]. Table 4.1 shows the primary details related to SOFC/mGT coupling.

The main drawback concerning SOFC/mGT coupling is related to the fact that the microturbines in such systems are not optimized for that particular application only, but rather, their designs are based on general market requirements in terms of size and performance. Thus, the adaptation of a commercial mGT for use in a fuel cell coupling application can introduce an important constraint on the stack size or can generate a significant efficiency reduction resulting from the need for an air bleed or bypass approach [11]. Strictly speaking, even if compensations can be provided for rated operation, fundamental performance degradation issues need to be tackled concerning off-design running [9]. Additional drawbacks relate to specific cycle considerations and/or involving the recuperated configuration. In detail, the particular efficiency of these systems is significantly affected by ambient temperature [12] (as in standard turbines), since a temperature increase brings about a higher power consumption in the compressor along with a lower mass flow rate through the engine. Moreover, an ambient temperature increase also generates a pressure ratio decrease that produces significant efficiency decay for both the SOFC and the turbine (also shown by an exhaust temperature increase). Additionally, care should be taken with exhaust temperature inasmuch as it cannot be decreased to values lower than 200–250°C (however, this disadvantage could be a beneficial aspect in the case of cogeneration for heating buildings or for producing steam for industrial processes) [3]. Finally, disadvantages related to dynamic and control aspects are present. The constraints that must be considered to mitigate or minimize some of the problems outlined above are discussed in the next section.

4.4 Constraints Related to SOFC/mGT Coupling

As already mentioned, SOFC/mGT coupling is not trivial, as shown by the various general plant layouts considered. Several additional property constraints have to be taken into account to avoid component damage or significant lifespan degradation [12–15]. The following factors have a strong influence on these constraints: plant layout, component layout and constraints, property ranges and controlling devices. A detailed constraint discussion is included below for the fuel cell, the microturbine and the control system.

- *Constraints related to the plant layout.* Since different plant layouts have been (and are) under analysis and development for SOFC-based hybrid systems, it is important to state that a specific plant configuration has to be adopted not only on the basis of performance and/or cost considerations, but also to take into account realistic design constraints. For instance, while atmospheric systems would appear to be more flexible regarding mechanical stresses in the cell and result in easier air management (the air flow rate control could simply be implemented via a bypass approach), the heat exchanger to be installed upstream of the expander would be an important issue for thermal considerations and would have an impact on both cost and reliability [16]. The introduction of flow recirculation (for both anodic or cathodic sides [17]) could be an interesting solution for flow pre-heating without external contributions (efficiency improvement), but would require attention to be given to operational constraints affecting the fuel cell in off-design and transient modes.

- *Constraints related to components.* Since each component has a specific layout and set of operational constraints, it is absolutely crucial to know all the relevant details for an optimized system design. For example, the fuel cell layout (tubular or planar) can significantly affect the entire system layout: while for a tubular cell, the stack can be directly connected to the recuperator outlet flange (the pre-heating tube is able to further increase airflow temperature), a planar stack needs a further external system (such as a recirculation device) to pre-heat the cathodic airflow upstream of the fuel cell [17]. Moreover, it is important to take into account the inherent component constraints (such as temperature, pressure, chemical composition, etc.) to avoid serious problems during operations. For instance, while a standard stainless steel recuperator is essential to exploit the exhaust thermal energy for air pre-heating up to 650°C [18], it cannot be used for higher temperature conditions (such as upstream of the expander in an atmospheric hybrid cycle based on high temperature SOFCs).

- *Constraints related to property ranges.* Since the operational conditions of the components can be quite different depending on the chosen technology (e.g. different types of SOFC operate at different temperatures), the relevant constraint issues have to be carefully analyzed. For instance, the choice of a low-temperature SOFC (in comparison with an SOFC operating close to 1000°C) can reduce the thermal operational constraint issues not only for the stack but also for some plant components including the mGT [2, 3, 19].

- *Constraints related to controlling devices.* Since the system design performance is not the only operational consideration, a realistic plant design has to incorporate optimized control elements for system dynamics and optimization (startup, shutdown and load change phases). It is also necessary to take into account possible constraints related to controlling devices [20]: valves and electrical systems (e.g. systems to control fuel cell current or mGT load). For instance, the valve technology (constraints and costs) has to be carefully taken into account during the detailed plant design as it affects the operational performance and plant costs. It is an important aspect of detailed system design, given that some valve installations are possible with standard technology (e.g. fuel, bleed or bypass valves), some solutions can be affected by high-cost issues for high-temperature conditions (such as a valve located downstream of the recuperator), while other installation types cannot be realistically managed (such as a valve installation at very high temperature conditions, as in the SOFC outlet flows) [21, 22].

4.4.1 Turbine System Constraints

On the turbine side, several constraints have to be carefully considered for both steady-state and transient conditions. While several research activities studied SOFC/mGT coupling from a thermodynamic point of view, considering layout optimization at steady-state conditions [23, 24], a good hybrid system development cannot neglect the substantial additional constraints regarding time-dependent operations. The following constraints specifically refer to the microturbine characteristics for optimum hybrid system performance.

- *Microturbine size.* Since commercially available microturbines are not specifically adopted for use in SOFC integrated hybrid system layouts, it is necessary to size the stack considering the mass flow values of available design specifications. A further consideration is the requirement for cell operation to be at a fixed temperature (or in a very narrow range) to maximize the inherent performance. Some research groups (such as the DLR [25]) have based their development activity on commercial designs and as such addressed these constraints. Solutions based on specifically designed micro-turbines have been pursued by various other groups (e.g. by Rolls-Royce Fuel Cell Systems [22]).

- *Compression ratio.* The correct selection of the microturbine compression ratio is essential. There is an undisputed beneficial influence of pressurization on the system cycle efficiency due to the nature of the cell electrochemical reaction. Practically, there is little flexibility in the compression ratio selection as standard machines equipped with a single-stage radial compressor are able to produce a compression ratio in the 4–5 range. To have higher compression ratios, it is necessary to design a multistage machine with a significant cost increase [22, 26, 27].

- *Microturbine layout.* The engine cycle layout is an essential constraint in hybrid system development. For instance, if the microturbine is not equipped with a recuperator, the air pre-heating has to be carried out with a significant cathode flow recirculation, while a recuperated machine can be connected directly to the fuel cell in the case of a tubular SOFC [17]. Since this aspect has an essential influence on plant costs, the constraints related to microturbine layout have to be carefully considered during the design process. Also, the layout of connecting pipes has a strong influence on system behaviour at both steady-state (additional pressure losses) and time-dependent operations (due to pipe volume).

- *Maximum turbine inlet temperature.* The maximum value of the turbine inlet temperature (TIT) is an essential constraint for appropriate matching with the fuel cell. This constraint will also influence the layout for the off-gas burner location (e.g. in case of a low TIT maximum value, it might be necessary to put the off-gas burner in a recirculation line instead of immediately upstream of the expander [17]). However, even if some concepts have been put forward, the radial expanders used in commercial microturbines are not designed using blade cooling technology. Therefore, no flexibility can be considered in increasing this parameter, because almost all commercial manufacturers produce machines with a TIT maximum value in the 900–950°C range [28]. More flexibility could be required in case of future activities, which would also involve extensive development of ceramic components [29] from thermal optimization considerations.

- *Machine load management at part-load conditions.* Machine load management at the part-load condition is essential for good cycle efficiency values. For instance, microturbine load management carried out at constant speed [30] is not the best solution from a system efficiency point of view, as (to maintain a constant SOFC temperature) the airflow rate reduction would have to be obtained with bleed or bypass operations. Such an approach would significantly affect the mGT efficiency as a large fraction of the airflow would be discharged after compression (bleed) or directly fed to the turbine at low-temperature conditions (bypass) [31]. A variable speed control approach (e.g. with a constant turbine outlet temperature control system) could be more efficient at part-load conditions [17]. However, variable speed operations have two main disadvantages: (i) pressure dynamics and pressure differences in the stack can be an issue because they can induce mechanical stresses; and (ii) compressor surge imposes limitations in terms of speed variation. Also, if the turbine speed is used as the actuator to maintain constant turbine outlet temperature (TOT), as stated, the use of an independent actuator such as a bypass valve for fuel cell thermal management is also necessary. On the other hand, if turbine speed is employed to regulate fuel cell temperature directly, limitations could occur regarding TOT. The utilization of variable geometry in the compressor or expander has also been proposed as means to control de-load and extend the range (turndown ratio) of a microturbine in a hybrid system.
- *Airflow management at part-load conditions.* Design and control considerations concerning the aforementioned constraint have to be integrated with the following aspects related to part-load conditions: since the airflow is directly dependent on the fuel cell load, no significant flexibility is present for the mGT. For this reason, some hybrid system designs are based on exclusive mGT controls to ensure a proper mass flow rate value (so the electrical grid connection is essential to compensate for possible power differences between demand and production). However, other

innovative approaches give the possibility of operation in a stand-alone mode (e.g. with a sharing out coefficient calculated to split the power between the generation systems [12, 20]).

- *Maximum rotational speed.* To avoid mechanical damage to the microturbine structure, mostly due to overspeeding of the rotor components, it is essential to consider the maximum sustainable rotational speed [32]. Even though this consideration is easily managed for steady-state running (because the maximum airflow rate corresponds to the mGT nominal speed to be reduced at part-load), possible risky conditions during transient operations have to be circumvented by the control system [12]. In comparison with a standard commercial mGT, the rotational speed control layout requires additional devices (such as a bypass valve [12]) as the expander is not located downstream of a fast-response system (such as the standard combustor), but receives the flow from a high capacitance device (the stack) with a very slow dynamic response. Additionally, a fundamental aspect of microturbines running at variable speed is the need to avoid operation near critical speeds, as this could cause an uncontrolled self-excitation of the shaft coupling with subsequent failure.

- *Minimum surge margin value.* Surge is an essential constraint for hybrid systems, given that the fuel cell components induce higher pressure losses between the compressor and expander, thus reducing the available surge margin [12, 20, 33]. Moreover, even if this constraint is satisfied in steady-state operation, it is still essential to take it into account during transient operation. Specifically, the coupling of the mGT with a large vessel containing SOFC components could generate critical transient response behaviours during startup, load changes or shutdown operations [12, 20]. For instance, if the mGT shutdown operation is carried out using the standard approach for commercial machines, a large size vessel (such as the fuel cell stack) would experience a slower response in depressurization in comparison with standard machines. So, this aspect would lead to surge operations in the case where this slow response is not sufficiently taken into account by the control system [34, 35].

4.4.2 SOFC System Constraints

The electrochemical reactor and the auxiliary components are important items concerning design constraints. Such issues arise mainly due to the various electrochemical reactions involving these components (not only the electrochemistry, but also all the reactions related to fuel processing) and to the high temperature of operation therein. Hence, the following primary constraints related to the fuel cell system have to be carefully observed in both steady-state and transient conditions.

- *Maximum temperature of the fuel cell, reformer and off-gas burner*. During various cell operations, it is important to avoid unusually high temperatures [12, 20]. It is essential to preserve component life and minimize performance degradation. However, the most critical aspects related to such considerations involve the complex SOFC layout. Specifically, since the stack and the reformer comprise several cells, it is necessary to avoid unusual temperature peaks, because damage in a cell due to over-temperature conditions generates significant maintenance costs [36]. So, a significant uniformity level in the flow and chemical composition distribution is essential to avoid such costly temperature peaks. Finally, regarding the off-gas burner design, it is essential to take into account the temperature limits of catalytic materials [37].

- *Maximum temperature gradients in ceramic components*. Since ceramic materials are very sensitive to temperature-induced stresses due to their brittle behaviour, it is necessary to avoid excessive thermal gradients between the different parts of the stacks [12]. For this reason, a well-controlled uniform flow and electrical current distribution is essential not only for steady-state conditions, but also during transients. It is also necessary to take into account the thermal expansion coefficient differences between the ceramic materials and alloy components, including high-temperature expansion devices where necessary. Furthermore, time-dependent thermal gradients are also key for thermal expansion considerations, hence it is necessary to avoid rapid temperature changes during all operational phases. Since the gradient limit is in the range of 3 K/min [20, 34], this is an overriding constraint responsible for the very long startup and shutdown phases [34].

- *High temperature issues for components.* The very high temperature conditions force the designers to use expensive temperature-resistant alloys for the stack vessel and the connection pipes. However, such constraints can be appropriately mitigated with the adoption of low-temperature SOFCs [38]. These latter devices would, at the same time, alleviate some of the temperature-related problems in the gas turbine (for instance, the recuperator), although at the expense of lower cycle efficiency.

- *Minimum steam-to-carbon value.* To avoid carbon deposition from hydrocarbon cracking (or other chemical reactions) at high temperature conditions, it is essential to maintain a large inventory of steam on the anodic side. Usually, the ratio between the steam and the hydrocarbon mole numbers (on the basis of carbon content) has to be higher than 1.8–2.0 [20]. This constraint implies a significant amount of steam injection into the anodic side from an external source or (more efficiently) the provision of a good anodic recirculation system [12, 20, 39].

- *Chemical composition and kinetic aspects.* Another important constraint relates to chemical composition and kinetics-related aspects, since the efficient stack operation requires maintaining flow compositions within very narrow limits [40, 41]. For instance, even if methane can be directly used as a fuel in SOFCs, the electrokinetic considerations regarding its consumption reveal a slow reaction rate, meaning that maintaining a significant amount of hydrogen in the anode is always necessary (to be produced by steam reforming and shifting reactions).

- *Mechanical constraints (e.g. maximum pressure or cathode/ anode pressure difference).* Design considerations related to mechanical strength impose conditions in order to avoid excessive stress on the components. While reasonable pressurization levels are easily sustained without limiting the material selection significantly (if the materials are correctly chosen to sustain high temperature conditions), the constraint related to the cathode/anode differential pressure is more critical [17]. Specifically, even if an appropriate pressure loss design is adopted to avoid significant pressure differentials at steady-state conditions [17], the different time-dependent responses of the fuel cell compartments (volume values and mass flow rate changes can be significantly different for the stack sides) can generate significant differential pressure peaks. Therefore, the mechanical design of the stack and the design of the control system must account for these transient operations/effects [12, 20].

4.4.3 Control System Constraints

Concluding the description of hybrid plant constraints, it is also necessary to consider control systems. Due to the very high plant complexity and the aforementioned constraints, a significant research effort has to be devoted to system dynamics and to the appropriate design of adequate control hardware and algorithms. Hence, the following brief outline of considerations regarding the specific constraints inherent to control system design is provided for both steady-state and transient operating conditions.

- *Commercially available control systems have to be redesigned.* Constraints outlined in the above sections mean that commercial mGT control systems cannot simply be transferred for use in the hybrid system. Hence the hardware has to be controlled via an integrated system for the whole plant [20, 34]. Usually, no mGT commercial control devices or approaches are commercially available for hybrid plants.
- *Avoiding high-temperature valves.* Since the plant comprises several lines to be managed, the simplest approach would involve the installation of several valves in the high-temperature ducts (e.g. in the recirculation ducts or between the stack and the turbine). However, this approach is not ideal given both cost and reliability issues; thus it is necessary to avoid valve installations in the high-temperature ducts. Controlling the required flows from the low-temperature zones causes a significant complexity increase in the control system design [12, 20].
- *Some critical properties cannot be measured.* Unfortunately, some control-specific processes and/or component variables that require careful monitoring cannot be easily measured

without excessive costs (e.g. since the flow chemical composition measurement requires very expensive devices, this property cannot be measured for control issues) [12, 20, 42]. Hence, since no measurements are carried out for these crucial variables, extensive simulations (with validated models) have to be carried out for control system design.

- *Simple PID controllers are not adequate to satisfy all system operational constraints.* Due to the slow thermal response of the stack and given the constraints related to temperature gradients, simple proportional integral derivative (PID) controllers are not able to completely avoid the oscillations of temperature during transient operations (e.g. load changes). Hence, alternative solutions have to be designed for one or more controllers considering the following approaches: feed-forward [20], model predictive control [43], h-infinity or other innovative control solutions [44].

4.5 Design and Off-design Aspects

Since several studies have taken place on the performance of hybrid systems under both design and off-design conditions, the main aspects related to these topics are discussed below.

4.5.1 Design Aspects

As previously stated, the initial efforts concerning hybrid system design and development were solely based on thermodynamic/electrochemical calculations [3]. Subsequent research activities were based on large scale SOFCs (to be coupled with conventional power plants) [3], notwithstanding the fact that, currently, most manufacturers are not able to scale-up SOFCs sufficiently for this approach to be practical. In these previous studies, very high efficiency values were obtained, with some optional outlays promising to operate close to 80% cycle efficiency [3]. An example of these preliminary calculations for the SOFC-based hybrid systems was presented in [3], with an electrical efficiency value of the order of 82%. Such a high value is primarily due to a large-scale gas turbine, low current density (and overly expensive) stacks and a very high recuperator inlet temperature (not sustainable by standard stainless steel components).

More commercially realistic hybrid system performance results were obtained in several studies based on the coupling between an SOFC and a small-scale turbine (a microturbine) which gave consideration to the component constraints (avoiding too exotic and expensive materials). For instance, a design efficiency value close to 70% was presented [45–48] for the top-performing layouts. Among these studies, several different plant configurations were analysed, starting from an atmospheric plant scheme [45] to different pressurized layouts, such as the coupling between an SOFC with a recuperated [45] or intercooled [46] machine. Due to the optimal performance being obtained from the recuperated mGT coupled with a pressurized SOFC, further investigation activities were carried out particularly on this layout.

Further design analyses were carried out to develop more realistic and feasible hybrid systems considering existing technology [49]. So, performance aspects for existing turbines [50, 51] were included to calculate plant operational conditions. Furthermore, to obtain realistic engineering results and hence valuable conclusions for the design, the power consumption related to auxiliaries and the thermal losses (for both components and pipes) were included in the calculations. Via these detailed models, realistic efficiency values close to 65% were calculated for small-size (hundreds of kilowatts) hybrid power generation systems. These works also highlight the beneficial influence of system pressurization. However, due to a more significant SOFC voltage increase in the 1–6 bar range [50] in comparison with the voltage increase that can be obtained with further pressurization, the hybrid system based on a pressurized SOFC in the 4–6 bar range was considered to be the best compromise between efficiency and costs. This aspect is also emphasized by the fact that commercial microturbines are able to produce a pressure ratio of 4 to 5 [28] without considering additional sophisticated and expensive solutions (e.g. recompression systems [26]). Further calculations including CO_2 sequestration systems show plant performance in the 60–62%

efficiency range [52, 53] due to the energy consumed by the CO_2 separation plant.

Many studies oriented towards prototype development [22, 54], maintain a focus on cost reduction, and as a result place less emphasis on efficiency maximization. For such reasons, the following technologies have been taken into account:

- flattened-tube or planar SOFCs to reduce manufacturing costs [22];
- plant layout simplifications to reduce expensive components (e.g. cathodic recirculation based on single-stage ejectors to avoid recuperator cost [22]);
- low-temperature SOFCs to reduce material costs [55];
- SOFC pressurization via a turbocharger instead of a micro-turbine to decrease the costs and the system size [54] (even if this layout is labelled as a hybrid system, it is just a pressurized SOFC, due to just one power generation device being included).

All these factors are responsible for a significant efficiency reduction to values below 60%. This performance, however, could prove to be acceptable if the plant cost reductions could compensate for the cost increase related to fuel consumption.

4.5.2 Off-design Aspects

The off-design analysis is essential for understanding the influence of different parameters on system performance at part-load conditions. Due to the critical behaviour at extreme low-load conditions related to the fuel cell, the analysis was usually carried out considering the plant load reduced to about 50% of the nominal value [47, 49, 56]. However, other studies (e.g. [53]) were extended to lower values on purely theoretical grounds (currently not considered feasible in realistic plant layouts).

Some initial works [46, 47] carried out regarding the off-design analysis of SOFC hybrid systems considered the variation of several critical system parameters (such as pressure ratio, combustor outlet temperature, machine efficiency values, moisture content in the fuel, etc.) and the plant layouts. However, the full set of real constraints were incorporated into the studies. For instance, in [47], results related to the operation with SOFC inlet temperature lower than 500°C were reported without the introduction of specific heating systems (e.g. a cathode recirculation).

Further studies [49, 56–58] based on more realistic cases considering all the steady-state constraints were carried out, considering microturbines able to operate at variable speeds; thus it is not possible to change the machine parameters (such as pressure ratio, mass flow rate, etc.) independently because they are calculated on the basis of compressor and turbine performance maps. Figure 4.5 [49] shows the plant net efficiency values obtained for a hybrid system including a tubular SOFC, an anodic recirculation device based on a single-stage ejector (located upstream of a pre-reformer), an off-gas burner located downstream of the stack, and a recuperated microturbine. The off-design efficiency trend (carried out at constant fuel utilization factor) shows a significant efficiency reduction when reducing the stack load at constant rotational speed. This is due to the fact that the fuel cell temperatures decrease with a load decrease if the air mass flow rate is maintained almost

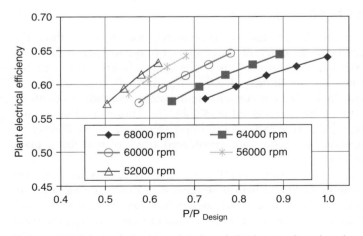

Figure 4.5 Off-design behaviour related to a hybrid system based on the coupling of a recuperated microturbine with a tubular SOFC.

constant (aiming to keep a constant machine speed). However, by reducing the plant load and the turbine rotational speed, it is possible to retain high efficiency also during part-load operations. This is a different behaviour in comparison with the traditional power generation systems, as it is now possible to operate the hybrid plants at high efficiency not only during design operations, but also at part-load conditions. So, this example of theoretical calculations on off-design performance of hybrid systems shows an important advantage for such plants for 'flexible' operations and applications related to the distributed generation approach.

4.6 Issues Related to Dynamic Aspects

To develop real hybrid system prototypes, the routinely adopted engineering steady-state analysis is not adequate, even if carried out at both design and off-design conditions. For instance, the plant must start operation from a startup phase that is very critical from a dynamic point of view. Load changes and other time-dependent operations have to be carefully analyzed to avoid critical conditions.

While the dynamic aspects of traditional plants (such as gas turbines) are well known and well managed by their commercial control tools, hybrid systems are not completely ready for market deployment owing to some issues concerning time-dependent operation. This is mainly due to high system complexity, coupled with a very large number of constraints related to components and/or parameters. As stated in the previous paragraphs, the coupling of a microturbine with an SOFC generates matching issues between a fast-acting system (the machine) and a slow-response component (the fuel cell) for the high thermal capacitance of the stack and the reformer. Moreover, while commercial microturbines are usually operated with a small volume size (pipes and combustor), the SOFC stack introduces a large volume connected between the compressor and the turbine (in pressurized systems), which has a significant impact on the dynamic performance of the machine. Usually, this addition causes an increased surge risk (especially during dynamic operations) in hybrid plants.

Although several studies [20, 21, 59–61] analyzed dynamic and control issues of hybrid systems carefully, a final solution is not available considering the large number of constraints and cost-related aspects already mentioned (dynamic issues can have a significant impact on costs in the case of a negative effect on plant life or on system efficiency). While some papers [20, 60, 61] presented control systems for hybrid plants based on SOFCs, the thermal stress problem for the fuel cell materials was not always considered (high thermal gradients over 3 K/min [62] are usually not allowed) or measurements were based on expensive probes (e.g. mass flow rate meters [20, 60]). Furthermore, the risks related to cathode/anode pressure difference increase were only carefully analyzed in a small number of studies [12, 20], while other dynamic works including control system development neglected these important constraints [17]. Time-dependent aspects of the anodic recirculation were neglected in several works based on constant recirculation ratios [62], or too simplistic solutions were utilized [61]. Due to the anodic circuit response aspects (e.g. a low steam-to-carbon ratio in the anodic circuit could cause harm to the entire plant), it is essential to perform calculations with validated dynamic tools (e.g. anodic ejector software [20, 63]) with the aim of studying plant control systems [20].

Since the dynamic aspects are linked to control system issues, matching slow- and fast-response components (SOFC and mGT devices) is a significant issue in the control system development. Several authors [20, 60, 61] have proposed control approaches based completely on PID devices, although further studies [12, 30, 44] showed that this solution has to be improved with further performance advances in controlling devices to satisfy all the system constraints. For instance, a simple PI or PID-based controller is not able to avoid SOFC temperature oscillations during load steps with thermal gradients higher than 3 K/min. So, coupling with advanced control approaches also needs to be considered, such as feed-forward [12], model predictive control (MPC) [30], or other techniques.

To give an example of the development of a full control system, it is interesting to discuss the main results obtained in [20], a study that considered a 300 kW hybrid plant based on the coupling of a tubular SOFC with a recuperated microturbine. As shown in [20], the machine speed is controlled by a bypass valve able to directly connect the compressor outlet with the turbine inlet ducts, and the plant is able to operate in stand-alone mode thanks to a battery system and a power-sharing coefficient (necessary to change the power demand distribution between the SOFC and the turbine during off-design operations). The dynamic operation was investigated by considering a plant load step decrease (−10%) starting from design operations. A comparison between a simple PI-based control system and an advanced approach equipped with an additional feed-forward tool was carried out with a Matlab-Simulink model to show the related performance. Due to the aforementioned dynamic aspects, the PI approach (despite allowing stable operations) is not able to satisfy the thermal gradient constraint, while the coupling with the feed-forward tools seem to be a good compromise between simplicity and safety (Figure 4.6). Moreover, Figure 4.7 shows that the feed-forward based control system is able to reduce the oscillations and the peaks related to the steam-to-carbon ratio (STCR) and cathode–anode pressure differences reducing stress on both chemical and mechanical sides.

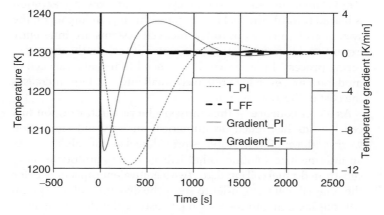

Figure 4.6 Dynamic and control system aspects related to a hybrid plant, based on the coupling of a recuperated microturbine with a tubular SOFC: fuel cell average temperature and thermal gradient.

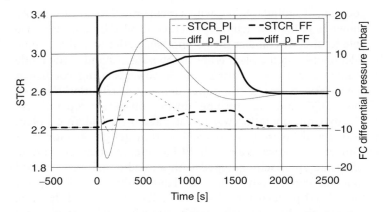

Figure 4.7 Dynamic and control system aspects related to a hybrid plant, based on the coupling of a recuperated microturbine with a tubular SOFC: steam-to-carbon ratio (STCR) and differential pressure.

4.7 Main Prototypes Developed for SOFC Hybrid Systems

The power ratio in a hybrid SOFC and micro gas turbine system is largely dependent on the cycle layout. Power generation is biased towards the fuel cell system. This has an impact on the system cost since, as of today, fuel cell systems are inherently expensive due to the lack of economies of scale in the manufacturing process. In other words, the need to install a fuel cell of moderate size, even for a relatively small microturbine, increases the cost of the system dramatically.

As a consequence of this, most of the research work on fuel cell hybrids has relied on numerical analysis and simulation work, with very few prototypes providing data for validation. As an intermediate solution, other hybrid system prototypes have substituted a fuel cell emulator (with equivalent capacitance) for the original SOFC, in order to study the dynamics of the system without incurring in the very high costs of the electrochemical components.

The paragraphs below present the most relevant experimental prototypes of SOFC-based hybrid systems so far.

4.7.1 Prototype by Siemens-Westinghouse

Siemens-Westinghouse pioneered the development of fuel cell hybrid systems in 1997 when the company received an order from Southern California Edison to design, build and test the first commercial power generation system, comprising a pressurized fuel cell and bottoming gas turbine [64]. The system was designed and built upon the company's experience with a 100 kWe atmospheric SOFC in a combined heat and power application in the Netherlands, which had already accumulated nearly 6000 hours achieving an average fuel-to-electric efficiency of 46% (75% total energy efficiency when considering hot water production) [65, 66]. This system was later refurbished and shipped to Torino, Italy, where it accumulated more than 27,000 hours by 2006 with 99.5% reliability [67].

The hybrid system deployed by the company in 1999 was based on a standard layout. A centrifugal compressor supplies air at approximately 3 bar to the fuel cell cathode, whilst the anode receives sulfur-free natural gas, which is later internally reformed, and a stream of recirculated anode exhaust gas. The remaining exhaust gases (from anode and cathode) are conveyed to an off-gas burner where the existing hydrogen (and, if necessary, additional natural gas) is burnt prior to entering the high-pressure turbine. This first turbine drives the compressor at high speed and makes use of a fraction of the available enthalpy content at the combustor outlet. The remaining energy is used to drive a second, low-speed expander that drives the generator.

The two-shaft micro gas turbine is based on a 75 kWe Ingersoll-Rand engine, whilst the SOFC is actually the same model used in the Dutch/Danish EDB/ESLAM consortium facility. This fuel cell comprises 48 bundle rows with a total of 1152 tubular cell stacks, with 1.5 m active length each; the whole assembly is enclosed in a high-pressure vessel for safety and mechanical integrity. The SOFC produces 100 kWe in atmospheric operation, but thanks to the pressurized concept used in the hybrid system, the output rises to 180 kWe (187 kWe DC power/176 kWe AC power); it is worth noting that, according to the company, no special measures were taken to adapt the cell to this high pressure [65]. The generator in the gas turbine produces some 47 kWe (AC power). The rated performance of the system is summarized in Table 4.2 below.

Table 4.2 Siemens-Westinghouse 220 kWe hybrid system performance.

Parameter	Rated value
Cell current	267 A
Cell voltage	0.610 V
Compressor pressure ratio	2.9:1
Air mass flow rate (compressor)	0.58 kg/s
Turbine inlet temperature	780°C
SOFC DC power	187 kWe
SOFC gross AC power	176 kWe
Turbogenerator AC power	47 kWe
Total net AC power	217 kWe
Net fuel-electric efficiency	57%

The system was installed at the National Fuel Cell Research Center (NFCRC), University of California Irvine, where the acceptance test took place in May 2000. The system ran for about 700 hours in the first year and experienced some technical problems regarding both electrical and thermal management issues. Amongst these, it is worth noting the very long time needed for the fuel cell to achieve full capacity (about 1000 hours) [68].

This early period was hence far from the objective of achieving 3000 operating hours, but without any doubt demonstrated the feasibility and reliability of the concept in spite of using components not developed specifically for the application (for instance, the micro gas turbine was found to be oversized). More recent works published by NFCRC have reported that the system was finally able to accumulate 2900 hours without major issues, achieving peak efficiencies of the order of 53%. This value is lower than the original objective, but still sets a milestone in the development of hybrid SOFC/GT systems [69].

As a final remark, the 220 kWe system was aimed at demonstrating the technology, with plans to scale it up to firstly to 300 kWe, based on an Allied Signal Parallon gas turbine [68], and later to a multi-megawatt system. This latter concept was designed around a 4.5 MWe Solar Mercury 50 engine coupled to an 8 MWe SOFC, which would theoretically yield overall electric efficiencies of the order of 60%. The total calculated output of the design (12.5 MWe) was selected according to an economic optimization process presented in [70]. Unfortunately, neither of these two systems was realized.

4.7.2 Prototype by Mitsubishi Heavy Industries

The activity of Mitsubishi Hitachi Power Systems (MHPS) in the field of solid oxide fuel cells and hybrid systems is much more recent than that of Siemens-Westinghouse. Nevertheless, the company has been able to catch up, demonstrating their first-generation system in the period from 2012 to 2014 and deploying a second-generation hybrid system more recently in 2015.

The hybrid system by MHPS is built for high power density, using a tubular fuel cell stack to take full advantage of the unmatched pressure resistance characteristics of this layout. As opposed to planar cells, sealing between cathode and anode is natural and does not require the exertion of mechanical forces normal to the stack. This is the same concept used by Siemens-Westinghouse previously, but in this case the main difference is that Siemens-Westinghouse employs a cathode-supported layout (with the anode on the outer side of the tube and the cathode on the inside), whereas MHPS utilizes a supporting tube with the cathode on the outer layer of the assembly and the anode in contact with the substrate, to enable lower voltage losses.

The first-generation system was delivered in 2013 and installed at the Senju Station of Tokyo Gas Ltd. It was based on the Model 10 stack, whose performance had already been demonstrated at 15 bar relative pressure and 900°C for more than 5000 h, and a recuperated micro gas turbine supplied by Toyota Turbine and Systems. In this layout, the compressor would supply air at 3 bar relative pressure to the cathode of the fuel cell and the expander would receive combustion gases at essentially the same pressure from the off-gas burner. The fuel cell assembly was enclosed in a high-pressure vessel to reduce the mechanical duty of the fuel cell and the heat losses to the surroundings, whereas the micro turbine was out of the vessel to ease accessibility [71].

The demonstrated voltage increase provided by high-pressure operation was between 5% and 10%, for 10 and 15 bar relative pressure respectively (with respect to operation at about 2.5 bar relative pressure). This pressure was largely reduced in the final hybrid system due to compression ratio limitations on the gas turbine compressor. The system was operated for 4000 h, achieving 206 kWe (183 kWe from the fuel cell and 23 kWe from the microturbine) and 50.2% low heating value (LHV) efficiency, which, even if far from the rated specifications (250 kWe and 55% efficiency), constituted a milestone in hybrid system development [72].

A second-generation system has recently been installed at Kyushu University, Japan, where it has been integrated into a smart grid providing combined heat and power to the campus [73, 74]. This new system implements two major innovations: a new stack (Model 15) that provides 50% higher power density, and a more compact packaging of the cells (i.e. an arrangement of cell bundles). Both features combined bring about a substantial reduction in the system footprint (40%) [75]. The system at Kyushu University has accumulated more than 3200 hours at constant output (200 kWe) with only minor shutdown events so far [73, 75]. Future plans include operation at peak power (250 kWe).

4.7.3 Prototype by Rolls-Royce Fuel Cell Systems

Rolls-Royce became interested in fuel cell systems in the late 1980s and started regular research activities in the first half of the next decade [76]. Then, after two decades of strong research activities and ties with universities worldwide, the majority of Rolls-Royce Fuel Cell Systems was acquired by LG Corporation, establishing LG Fuel Cell Systems Inc. in 2012. The assets and facilities were also taken over by LG, hence incurring minimum interruption of the activities developed by the company.

The Rolls-Royce Fuel Cell Systems company has built around their proprietary integrated planar design, which is a segmented tubular configuration claimed to unite the best features of tubular and planar designs: management of thermal expansion and shortest electric current paths (thus lower ohmic losses) [76]. A detailed discussion about the performance of this technology is presented elsewhere [58, 77]. The system makes use of a similar concept to Siemens-Westinghouse and Mitsubishi Hitachi Power Systems. A compressor supplies pressurized air to the fuel cell cathode, whereas partially reformed fuel is supplied to the anode. The outflows from both electrodes are mixed and burned in an off-gas burner prior to being expanded in a centripetal turbine. In comparison with the previous prototype, this system includes cathodic recirculation (operated by a single-stage ejector) instead of a recuperator. This approach reduces plant costs by removing this heat exchanger.

The University of Genoa (as Rolls-Royce University Technology Centre) has been involved in modelling activities on this plant since 2004. Several analyses were carried out on this prototype starting from design point activities to off-design and transient calculations (including the identification of control system problems and solutions). For instance, a detailed off-design analysis was reported in [56] studying the effect of ambient temperature variation. Another important published activity on this plant related the start-up phase calculation to a real-time tool [78].

Unfortunately, no further performance data are available for this prototype and, at the moment, it is not possible to write the objectives reached with this research activity. Only the experimental information produced by hybrid system emulators has been made available in the public domain [22, 79]. These emulators [80] are actually hybrid systems where the fuel cell system is substituted by a volume with equivalent capacitance, hence replicating the dynamic performance of the gas turbine as implemented in a practical hybrid system. This is very important to characterize the safe performance map, avoiding regions that can potentially run the system into surge.

4.8 Conclusions

Hybrid systems can be looked at from two different angles. On one hand, they can be considered as means to recuperate waste heat effectively from a high-temperature fuel cell, thus raising the overall efficiency of the power generation system (also by potentially increasing the operating pressure of the fuel cell, with the corresponding performance enhancement). On the other hand, hybrid systems can also be regarded as heat engines where heat addition is accompanied by the production of electrochemical work. The concept can be materialized in a number of layouts wherein the fuel cell is coupled with an open or closed gas turbine, steam turbine or even Stirling engine. Whichever is the case, these hybrids enable a step change in power generation efficiency with respect to existing technology.

The combination of subsystems with such different natures (fuel cells and heat engines) poses a number of technical challenges that manifest as constraints for the hybrid system. This chapter has discussed some of them. For instance, special care must be taken with the thermo-mechanical integration of cell components, given the risk of mechanical failure coming from thermal expansion mismatch in the brittle fuel cell elements. This, in turn, sets important restrictions on the control system, which must ensure that temperature gradients are below certain threshold values. The selection of appropriate control elements like valves and similar is key to achieving this objective.

Regarding the microturbine, the main hurdle is the lack of engines specifically developed for integration in hybrid systems, which necessitates the adaption of engines whose characteristics do not fit with the boundary conditions of the system of interest. Indeed, all the engines available on the market are designed for fossil fuel combustion, with just some of them adapted to other heat sources (for instance solar). The introduction of a fuel cell between the compressor and expander brings about a modification of the stationary and transient operation of the unit. Accordingly, a number of features of the engine must be double-checked for safety. First, the turbine inlet temperature is likely to increase as a consequence of combustion in the off-gas burner; nevertheless, it must typically be kept below 900–950°C in radial expanders to ensure mechanical integrity. Also, the added volume (due to the SOFC) between turbomachinery components is likely to reduce the surge margin, thus increasing the risk of unstable operation during transient manoeuvres.

Additional constraints arise when the integration is considered, such as the need to control the steam-to-carbon ratio in the cell via recirculation flows and also the integration of fuel reforming into the energy balance of the system. This increases the complexity of the control system, which at this stage cannot make use of simple PID controllers as discussed earlier. More complex approaches are thus needed which, if successfully developed, can provide improved part-load performance over a wider load range.

The challenge related to hybrid system development is difficult, as deduced from the small number of demonstration systems that have been operated to date. It should be expected, though, that vital information obtained through experimental activities will continue to shed light on methods to ease the difficult integration of fuel cells and micro gas turbines. So far, the cumulative experience of all these field units amounts to more than 10,000 hours and has proved vital in identifying, studying and removing technical difficulties found in fuel cells, microturbines and integrated systems.

4.9 Questions and Exercises

1 What are the main constraints related to SOFC components for developing a hybrid system?

2 What is the efficiency range of hybrid systems based on pressurized SOFCs?

3 Why have calculated design efficiency values decreased from initial calculations to those found in the latest design activities?

4 A hybrid system is based on the coupling of an SOFC with a recuperated mGT, as shown in Figure 4.3. Considering air in both compressor and turbine components ($k = 1.4$) and neglecting pressure losses, what is the hybrid system efficiency value (all the data are given in Table 4.3)?

5 Considering the previous exercise, if the fuel cell inlet temperature is equal to $620°C$ on the cathodic side, what is the recuperator effectiveness?

6 What variable decrease is necessary to maintain a high efficiency level during off-design conditions?

Table 4.3 Data for Question 4.

Parameter	Value
mGT compression ratio	4.0
SOFC power (AC)	240 kW
OGB outlet temperature	1000°C
Compressor mass flow rate	0.65 kg/s
Fuel mass flow rate	0.01 kg/s
Fuel LHV	47 MJ/kg
Specific heat ratio (air)	1.4
mGT mechanical efficiency	0.99
mGT electrical efficiency (generator and power conditioning system)	0.88
Compressor isentropic efficiency	0.78
Turbine isentropic efficiency	0.85

7   Why would PI/PID control approaches alone be unable to satisfy all the system constraints during time-dependent operations?

8   Cite three areas of concern when integrating a solid oxide fuel cell and microturbine into a hybrid system.

9   What is the usual share of power generation (fuel cell and microturbine) in a standard hybrid system?

References

1 Otomo, J., Oishi, J., Mitsumori, T., Iwasaki, H. and Yamada K. (2013) Evaluation of cost reduction potential for 1 kW class SOFC stack production: Implications for SOFC technology scenario. *International Journal of Hydrogen Energy*, 38, 14337–14347.
2 Larminie, J. and Dicks, A. (2003) *Fuel Cell Systems Explained* (2nd edn). John Wiley & Sons Ltd, Chichester.
3 EG&G Technical Services, Inc. (2004) *Fuel Cell Handbook* (7th edn). EG&G Technical Services, Inc., Morgantown, WV.

4 Veyo, S.E., Lundberg, W.L., Vora, S.D. and Litzinger, K.P. (2003) Tubular SOFC hybrid power system status. ASME Turbo Expo 2003, Atlanta, GA.

5 Ugartemendia, J., Ostolaza, J.X. and Zubia, I. (2013) Operating point optimization of a hydrogen fueled hybrid solid oxide fuel cell-steam turbine (SOFC-ST) plant. *Energies*, 6, 5046–5068.

6 Nguyen, Q.M. (2005) Solid oxide fuel cell systems for stationary power generation applications. *Electrochemical Society Proceedings*, 7, 76–81.

7 Wilson, J.D., Colson, C.M. and Nehrir, M.H. (2010) Cost and unit-sizing analysis of a hybrid SOFC/microturbine generation system for residential applications. North American Power Symposium, Arlington, TX.

8 Duan, L., Huang, K., Zhang, X. and Yang, Y. (2013) Comparison study on different SOFC hybrid systems with zero-CO_2 emission. *Energy*, 58, 66–77.

9 Costamagna, P., Magistri, L. and Massardo, A.F. (2001) Design and part-load performance of a hybrid system based on a solid oxide fuel cell reactor and a micro gas turbine. *Journal of Power Sources*, 96, 352–368.

10 Ferrari, M.L., Traverso, A., Pascenti, M. and Massardo, A.F. (2007) Early start-up of SOFC hybrid systems with ejector cathodic recirculation: experimental results and model verification. *Proceedings of the Institution of Mechanical Engineers, Part A, Journal of Power and Energy*, 221, 627–635.

11 Bakalis, D.P. and Stamatis, A.G. (2014) Improving hybrid SOFC-GT systems performance through turbomachinery design. *International Journal of Energy Research*, 38, 1975–1986.

12 Ferrari, M.L. (2015) Advanced control approach for hybrid systems based on solid oxide fuel cells. *Applied Energy*, 145, 364-373.

13 Du, L.X., Ma, T., Zeng, M., Guo, Z.X. and Wang, Q.W. (2014) Numerical investigations on the thermohydraulic performance of cross-wavy channels with multi-periodic boundary conditions. *Numerical Heat Transfer; Part A: Applications*, 65, 732–749.

14 Massardo, A.F. and Bosio, B. (2000) Assessment of molten carbonate fuel cell models and integration with gas and steam cycles. *Journal of Engineering of Gas Turbine and Power*, 124, 103–109.

15 Zaccaria, V., Tucker, D. and Traverso, A. (2016) Transfer function development for SOFC/GT hybrid systems control using cold air bypass. *Applied Energy*, 165, 695–706.

16 Roberts, R., Brouwer, J., Jabbari, F., Junker, T. and Ghezel-Ayagh, H. (2006) Control design of an atmospheric solid oxide fuel cell/gas turbine hybrid system: Variable versus fixed speed gas turbine operation. *Journal of Power Sources*, 161, 484–491.

17 Ferrari, M.L. and Massardo, A.F. (2013) Cathode-anode interaction in SOFC hybrid systems. *Applied Energy*, 105, 369–379.

18 McDonald, C.F. (1999) Low-cost compact primary surface recuperator concept for microturbines. *Applied Thermal Engineering*, 20, 471–497.

19 Fan, L., Wang, C., Chen, M. and Zhu, B. (2013) Recent development of ceria-based (nano)composite materials for low temperature ceramic fuel cells and electrolyte-free fuel cells. *Journal of Power Sources*, 234, 154–174.

20 Ferrari, M.L. (2011) Solid oxide fuel cell hybrid system: control strategy for stand-alone configurations. *Journal of Power Sources*, 196, 2682–2690.

21 Stiller, C., Thoruda, B., Bolland, O., Kandepu, R. and Lars Imsland, L. (2006) Control strategy for a solid oxide fuel cell and gas turbine hybrid system. *Journal of Power Sources*, 158, 303–315.

22 Agnew, G.D., Bozzolo, M., Moritz, R.R. and Berenyi, S. (2005) *The Design and Integration of the Rolls-Royce Fuel Cell Systems 1MW SOFC*. ASME Paper GT2005-69122.

23 Massardo, A.F. and Magistri, L. (2013) Internal reforming solid oxide fuel cell gas turbine combined cycles (IRSOFC-GT). Part II: Energy and thermoeconomic analyses. *Journal of Engineering for Gas Turbines and Power*, 125, 67–74.

24 Song, T.W., Sohn, J.L., Kim, J.H., Kim, T.S., Ro, S.T. and Suzuki, K. (2005) Performance analysis of a tubular solid oxide fuel cell/micro gas turbine hybrid power system based on a quasi-two dimensional model. *Journal of Power Sources*, 142, 30–42.

25 Hohloch, M., Widenhorn, A., Lebküchner, D., Panne, T. and Aigner, M. (2008) Micro gas turbine test rig for hybrid power plant application. ASME Turbo Expo 2008, Paper GT2008-50443.

26 Damo, U.M., Ferrari, M.L., Turan, A. and Massardo, A.F. (2015) Re-compression model for SOFC hybrid systems: start-up and shutdown test for an emulator rig. *Fuel Cells*, 1, 42–48.

27 Damianos, T.-I., Kalfas, A.I., Ferrari, M.L., Massardo, A.F. and Magistri, L. (2013) Integrated design of centrifugal compressor and radial turbine of hybrid mGT/SOFC power train emulator. ISABE 2013, Paper no. 1707, Busan, Corea del Sud.

28 Hamilton, S.L. (2003) *The Handbook of Microturbine Generators*. Pennwell, Tulsa, Oklahoma.

29 Wu, H., Li, D. and Tang, Y. (2013) Fabrication of integral core/shell ceramic casting mould for hollow turbine blade. *Applied Mechanics and Materials*, 248, 231–236.

30 Caratozzolo, F., Ferrari, M.L., Traverso, A. and Massardo, A.F. (2013) Emulator rig for SOFC hybrid systems: temperature and power control with a real-time software. *Fuel Cells*, 6, 1123–1130.

31 Tsai, A., Tucker, D. and Clippinger, D. (2011) Simultaneous turbine speed regulation and fuel cell airflow tracking of a SOFC/GT hybrid plant with the use of airflow bypass valves. *Journal of Fuel Cell Science and Technology*, 8, 061018_1-10.

32 McLarty, D., Brouwer, J. and Samuelsen, S. (2014) Fuel cell gas turbine hybrid system design part II: Dynamics and control. *Journal of Power Sources*, 254, 126–136.

33 Tucker, D., Lawson, L., Gemmen, R. and Dennis, R. (2005) Evaluation of hybrid fuel cell turbine system startup with compressor bleed. ASME Turbo Expo 2005, Paper GT2008-68784.

34 Ferrari, M.L., Pascenti, M., Magistri, L. and Massardo, A.F. (2008) Emulation of hybrid system start-up and shutdown phases with a micro gas turbine based test rig. ASME Turbo Expo 2008, Paper GT2008-50617.

35 Houldershaw, M. (2017) Compact, efficient, low cost turbomachine compressor for automotive fuel cell systems. TSB Project 101580, Aeristech. Available at http://gtr.rcuk .ac.uk/projects?ref=101580

36 Richards, G.A., McMillian, M.M., Gemmen, R.S., Rogers, W.A. and Cully, S.R. (2001) Issues for low-emission, fuel-flexible power systems. *Progress in Energy and Combustion Science*, 27, 141–169.

37 Sudaprasert, K., Travis, R.P. and Martinez-Botas, R.F. (2010) A study of temperature distribution across a solid oxide fuel cell stack. *Journal of Fuel Cell Science and Technology*, 7 011002_1-13.

38 Patakangas, J., Ma, Y., Jing, Y. and Lund, P. (2014) Review and analysis of characterization methods and ionic conductivities for low-temperature solid oxide fuel cells (LT-SOFC). *Journal of Power Sources*, 263, 315–331.

39 Ferrari, M.L., Pascenti, M. and Massardo, A.F. (2008) Ejector model for high temperature fuel cell hybrid systems: experimental validation at steady-state and dynamic conditions. *Journal of Fuel Cell Science and Technology*, 5, 041005_1-7.

40 Harun, N.F., Tucker, D. and Adams, T.A. (2016) Impact of fuel composition transients on SOFC performance in gas turbine hybrid systems. *Applied Energy*, 164, 446–461.

41 Ferrari, M.L., Pascenti, M., Traverso, A.N. and Massardo, A.F. (2012) Hybrid system test rig: chemical composition emulation with steam injection. *Applied Energy*, 97, 809–815.

42 Nanaeda, K., Mueller, F., Brouwer, J. and Samuelsen, S. (2010) Dynamic modeling and evaluation of solid oxide fuel cell – combined heat and power system operating strategies. *Journal of Power Sources*, 195, 3176–3185.

43 Larosa, L., Traverso, A., Ferrari, M.L. and Zaccaria, V. (2015) Pressurized SOFC hybrid systems: control system study and experimental verification. *Journal of Engineering for Gas Turbines and Power*, 137, 031602_1-8.

44 Tsai, A., Tucker, D. and Emami, T. (2014) Adaptive control of a nonlinear fuel cell-gas turbine balance of plant simulation facility. *Journal of Fuel Cell Science and Technology*, 11, 061002_1-8.

45 Massardo, A.F. and Lubelli, F. (2000) Internal reforming solid oxide fuel cell – gas turbine combined cycles (IRSOFC-GT). Part A: Cell model and cycle thermodynamic analysis. *Journal of Engineering for Gas Turbines and Power*, 122, 27–35.

46 Yi, Y., Rao, A.D., Brouwer, J. and Samuelsen, G.S. (2004) Analysis and optimization of a solid oxide fuel cell and intercooled gas turbine (SOFC–ICGT) hybrid cycle. *Journal of Power Sources*, 132, 77–85.

47 Stephenson, D. and Ritchey, I. (1997) Parametric study of fuel cell and gas turbine combined cycle performance. ASME Paper 97-GT-340, International Gas Turbine & Aeroengine Congress & Exhibition, Orlando, Florida.

48 Achehnbach, E. (1994) Three-dimensional and time-dependent simulation of a planar SOFC stack. *Journal of Power Sources*, 49, 333–348.

49 Ferrari, M.L., Traverso, A., Magistri, L. and Massardo, A.F. (2005) Influence of the anodic recirculation transient behaviour on the SOFC hybrid system performance. *Journal of Power Sources*, 149, 22–32.

50 Liese, E.A., Ferrari, M.L., VanOsdol, J., Tucker, D. and Gemmen, R.S. (2008) Modeling of combined SOFC and turbine power systems, in *Modeling Solid Oxide Fuel Cells*, Springer Science Business Media, pp. 239–268.

51 McLarty, D., Brouwer, J. and Samuelsen, S. (2013) Hybrid fuel cell gas turbine system design and optimization. *Journal of Fuel Cell Science and Technology*, 10, 041005_1-11.

52 Spallina, V., Romano, M.C., Campanari, S. and Lozza, G. (2010) A SOFC-based integrated gasification fuel cell cycle with CO_2 capture. *Journal of Engineering for Gas Turbines and Power*, 132, 1–10.

53 Fredriksson Möller, B., Arriagada, J., Assadi, M. and Potts I. (2004) Optimisation of an SOFC/GT system with CO_2-capture. *Journal of Power Sources*, 131, 320–326.

54 Powered by D'Appolonia (2015) Available at http://www.bio-hypp.eu/

55 Liu, Q.L., Khor, K.A. and Chan, S.H. (2006) High-performance low-temperature solid oxide fuel cell with novel BSCF cathode. *Journal of Power Sources*, 161, 123–128.

56 Trasino, F., Bozzolo, M., Magistri, L. and Massardo, A.F. (2011) Modeling and performance analysis of the Rolls-Royce Fuel Cell Systems Limited: 1 MW plant. *Journal of Engineering for Gas Turbines and Power*, 133, 021701-1-11.

57 Choi, J., Sohn, J.L., Song, S.J. and Kim, T.S. (2010) Off-design performance characteristics of SOFC-GT hybrid system

operating with syngas fuel. *Transactions of the Korean Society of Mechanical Engineers*, 34, 269–274.

58 Magistri, L., Traverso, A., Cerutti, F., Bozzolo, M., Costamagna, P. and Massardo, A.F. (2005) Modelling of pressurised hybrid systems based on integrated planar solid oxide fuel cell (IP-SOFC) technology. *Fuel Cells*, 1, 80.96.

59 Hawkes, A.D., Aguiar, P., Croxford, B., Leach, M.A., Adjiman, C.S. and Brandon, N.P. (2007) Solid oxide fuel cell micro combined heat and power system operating strategy: Options for provision of residential space and water heating. *Journal of Power Sources*, 164, 260–271.

60 Henke, M., Willich, C., Westner, C., Leucht, F., Kallo, J., Bessler, W.G. and Friedrich, K.A. (2013) A validated multi-scale model of a SOFC stack at elevated pressure. *Fuel Cells*, 13, 773–780.

61 Jia, Z., Sun, J., Oh, S.-R., Dobbs, H. and King, J. (2013) Control of the dual mode operation of generator/motor in SOFC/GT-based APU for extended dynamic capabilities. *Journal of Power Sources*, 235, 172–180.

62 Ferrari, M.L., Pascenti, M., Magistri, L. and Massardo, A.F. (2010) Hybrid system test rig: Start-up and shutdown physical emulation. *Journal of Fuel Cell Science and Technology*, 7, 021005_1-7.

63 Rao, S.M.V. and Jagadeesh, G. (2014) Novel supersonic nozzles for mixing enhancement in supersonic ejectors. *Applied Thermal Engineering*, 71, 62–71.

64 George, R.A. (2000) Status of tubular SOFC field unit demonstrations. *Journal of Power Sources*, 86, 134–139.

65 Veyo, S.E., Shockling, L.A., Dederer, J.T., Gillet, J.E. and Lundberg, W.L. (2002) Tubular solid oxide fuel cell/gas turbine hybrid cycle power systems: Status. *Journal of Engineering for Gas Turbines and Power*, 124, 845–849.

66 Veyo, S.A. (1996) Westinghouse 100 kWe SOFC Demonstration Status. Fuel Cells '6 Review Meeting, Morgantown. Available at www.osti.gov/scitech/servlets/purl/462945

67 Calì, M., Leone, P., Santarelli, M., Orsello, G. and Disegna, G. (2006) Operation of the tubular SOFC CHP100 kWe Field Unit in Italy. General topics and operation description by means of regression models. World Hydrogen Energy Conference, Lyon.

68 Van Gerwen, R.J.F. (2003) *High Temperature Solid Oxide Fuel Cells: Fundamentals, Designs and Applications.* Elsevier, Oxford.

69 Roberts, R.A. and Brouwer, J. (2006) Dynamic simulation of a pressurized 220 kW solid oxide fuel-cell–gas-turbine hybrid system: Modeled performance compared to measured results. *Journal of Fuel Cell Science and Technology*, 3, 18–25.

70 Lundberg, W.L., Veyo, S.E. and Moeckel, M.D. (2003) A high-efficiency solid oxide fuel cell hybrid power system using the Mercury 50 Advanced Turbine Systems gas turbine. *Journal of Engineering for Gas Turbines and Power*, 125, 51–58.

71 Kobayashi, Y., Tomida, K., Nishiura, M., Hiwatashi, K., Kishizawa, H. and Takenobu, K. (2015) *Mitsubishi Heavy Industries Technical Review*, 52, 111–116.

72 Fuel Cells Bulletin (2014) *German coalition expanding hydrogen network.* Fuel Cells Bulletin, October, p. 4.

73 Ando, Y., Oozawa, H., Mihara, M., Irie, H., Urashita, Y. and Ikegami, T. (2015) *Mitsubishi Heavy Industries Technical Review*, 52, 47–52.

74 Fuel Cells Bulletin (2014) Mitsubishi Hitachi to integrate SOFC with micro gas turbine for Kyushu University demonstration. *Fuel Cells Bulletin*, December, p. 1.

75 Kitagawa, Y., Tomida, K., Nishiura, M., Hiwatashi, K., Kishizawa, H., Oozawa, H., Kobayashi, Y., Takeuchi, Y. and Mihara, M. (2015) Development of high efficiency SOFC power generation at MHPS. Fuel Cell Seminar and Energy Exposition, Westin Bonaventure.

76 Gardner, F.J., Day, M.J., Brandon, N.P., Pashley, M.N. and Cassidy, M. (2000) SOFC technology development at Rolls-Royce. *Journal of Power Sources*, 86, 122–129.

77 Costamagna, P., Grosso, S., Travis, R. and Magistri, L. (2015) Integrated planar solid oxide fuel cell: steady-state model of a bundle and validation through single tube experimental data. *Energies*, 8, 13231–13254.

78 Ghigliazza, F., Traverso, A., Massardo, A.F., Wingate, J. and Ferrari, M.L. (2009) Generic Real-Time Modeling of Solid Oxide Fuel Cell Hybrid Systems. *Journal of Fuel Cell Science and Technology*, 6, 021312_1-7.

79 Magistri, L., Bozzolo, M., Tarnowski, O., Agnew, G. and Massardo, A.A. (2007) Design and off-design analysis of a MW hybrid system based on Rolls-Royce integrated planar solid oxide fuel cells. *Journal of Fuel Cell Science and Technology*, 129, 792–797.

80 Larosa, L., Traverso, A. and Massardo, A.F. (2016) Dynamic analysis of a recuperated mGT cycle for fuel cell hybrid systems. ASME Paper GT2016-57312, ASME Turbo Expo 2016, Seoul, South Korea.

5

Computational Models for Hybrid Systems

CHAPTER OVERVIEW

5.1   Introduction

The design and analysis of complex hybrid systems, with special attention given to the issues and constraints outlined in Chapter 4, can be approached using a range of computational models. Several tools exist aimed towards the design, development and performance analysis of hybrid systems, including off-design and dynamic behaviour at both component and system levels. For component calculations, especially for fuel cells, very detailed approaches need to be employed to adequately capture the underlying physics. For instance, electrolyte selection, electrode design and the investigation/optimization of electrode performance require very complex chemical models involving micro-kinetic descriptions. Because of this, the fluid dynamic and transport aspects and relevant variable/property field distributions for SOFCs have been the focus of numerous studies in the literature, employing complex 3D computational fluid dynamics (CFD) tools coupled with detailed electrochemical

Hybrid Systems Based on Solid Oxide Fuel Cells: Modelling and Design, First Edition.
Mario L. Ferrari, Usman M. Damo, Ali Turan, and David Sánchez.

reactions. There are cases where it is either prudent or necessary to reduce the model complexity, such as in the case of symmetrical geometries (2D approaches) and or when a rapid numerical solution is required (1D approaches). At plant level, model simplification is essential in order to have reasonable calculation times. For this reason, macro-engineering models based on control volume analyses (zero-dimensional (0D) approaches) are generally used to obtain reasonable performance-level predictions including system dynamics. If real-time performance is required, the modelling approaches have to be extremely simplified, but must still be able to capture adequately the requested phenomena. These macro-engineering models are usually based on simple global equations or, where the physical aspects are too complicated, on interpolation between performance maps. Real-time modelling approaches are also considered essential for plant monitoring and diagnostic reasons, because a model running in parallel with the plant can be useful for fault detection.

It is of great importance that theoretical models have to be thoroughly validated to assess their reliability, especially where there are a large number of simplifications. Since experimental data related to the performance of entire hybrid systems are sorely lacking, model validation activities are usually carried out at the component level or by comparison with more reliable (and sometimes more complex) models.

Figure 5.1 presents a summary of different modelling approaches for hybrid systems using the conventional categories (white, grey and black boxes) to define the detail levels. This figure also shows the relative level of complexity and computing time required for each approach. The modelling techniques that may be applied to each of these systems can range from complex approaches (such as full equation models and 3D approaches) typically used for components or subcomponents, to the simplified approaches (1D, 0D or look-up table based techniques) for plant-level simulations.

A typical approach to analysing a hybrid system utilizes 0D techniques for all components (e.g. compressor, turbine, connecting pipes, combustor, etc.) except for the SOFC and recuperator. These two components could require 1D or quasi-2D

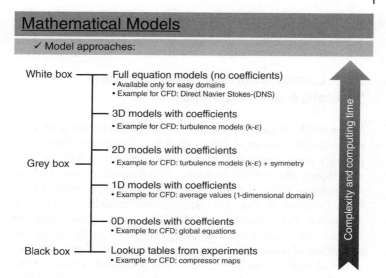

Figure 5.1 Different modelling approaches.

approaches depending on the different flows, in order to produce reliable results.

5.2 Steady-state Models for Hybrid Systems

A plant design activity starts with parametric cycle analysis. At this stage all design choices are still under control (i.e. not fixed) and the system size and numerous other property values are yet to be determined. While this stage is usually referred to as 'on-design' cycle analysis, 'off-design' studies deal with the behaviour of a defined system under conditions different from design values (e.g. part-load operations or ambient conditions different from those at the design point). The objective of these analyses (both on-design and off-design calculations) is to obtain an estimate of the performance parameters (such as generated power, efficiency values and the main system properties) taking into account the component constraints (such as maximum allowable turbine inlet temperature), the environmental conditions (ambient pressure, temperature and humidity), and design choices (such as the compressor pressure ratio).

5.3 Computational Models for Hybrid Systems: Modelling Steps

When modelling any hybrid system (HS) plant layout (a fuel cell coupled with a traditional system), it is necessary to consider the primary requirements of the model in order to determine suitable inputs and outputs and their corresponding ranges where appropriate. Once the main attributes and requirements have been established, the most appropriate modelling technique and detailed characteristics of the model can be determined to best suit the application. This important step usually introduces a number of assumptions, an unavoidable part of this process, that will give rise to inaccuracies and potentially influence the model to the extent where it can predict incorrect trends. More details about the model can be determined subsequently, after finalizing such criteria. As with the modelling of other thermal systems encountered in the literature, the modelling of an SOFC HS involves understanding the system and then translating it into mathematical equations and statements [1]. The common steps for model development are summarized as follows:

1) specifying a control volume around the desired system;
2) writing general laws (including conservation of mass, energy, and momentum; second law of thermodynamics; charge balance, etc.);
3) specifying boundary and initial conditions;
4) solving governing equations by considering boundary and initial conditions (through an analytical or numerical solution);
5) validating the model [1].

In spite of the fact that fuel cell simulation is a three-dimensional (3D) and time-dependent problem, using appropriate assumptions it can be reduced in complexity to a steady-state 2D, 1D, or 0D problem for different applications and objectives [1]. As discussed in previous chapters, most SOFC HS simulations in the open literature are 0D models. In this type of modelling, a series of mathematical formulations are utilized to define output variables based on input variables [1]. In this technique, the fuel cell is processed as a dimensionless box, hence the term employed by some authors as 'black box modelling' [1]. In spite of the large numbers of assumptions and simplifications in this method, it is a vital approach when examining the effects of various operational parameters on the cycle overall performance, performing a sensitivity analysis, and comparing different configurations. When the goal of the modelling is to study the inner workings of a SOFC, a 0D approach is not suitable. Nonetheless, for HS system simulation, where emphasis is placed on studying the interaction between the fuel cell and the rest of the system and how the fuel cell can affect the overall performance of the system, this approach is usually appropriate. In this level of system modelling, there are a variety of assumptions and simplifications made. For example, a simple modelling approach was adopted by Winkler *et al.* [2] who studied a hybrid fuel cell cycle. The main modelling assumptions used in this study were based on a fuel cell operated reversibly, and the coupled heat engine system was a Carnot cycle.

Various software packages and programming languages have been used in SOFC HS simulations. Commercial models for SOFC stacks are not yet available, and hence all modellers usually devise their own with proper details and assumptions [1]. What further differentiates the approaches these modellers take is how they simulate the remaining components of the system. In terms of categorizing the use or development of software used to model SOFC stacks, approaches can be grouped together into two primary classes. In the first category, comprehensive models can be developed using programming languages such as FORTRAN or high-level software such as a MATLAB/Simulink® platform to solve the governing equations for the system, while the second approach utilizes commercial software such as Aspen Plus® to model the conventional components of the cycle.

Due to the nature of the assumptions and simplifications involved with numerical modelling approaches, resulting predictions should be used cautiously. In every modelling activity, physical attributes of the system should be translated into mathematical equations and solutions of these equations are used to express the behaviour of the system [1]. In the case of fuel cells, the physical realities include highly complex physicochemical processes embodying subtle details. Therefore, in order to extract and subsequently solve these governing equations, a large number of assumptions and simplifications are usually employed for predictions of performance parameters and main properties (such as mass flow rate, pressure and temperature values). Fuel cell models are a 'simplified representation of real physics', and even with appropriate validation, the accuracy of their results over a range of conditions can be questionable, with oversimplification introducing significant errors [1].

An example of such problems is reported in the work of Bove *et al.* [3], which highlights that modelling using the standard 0D technique typically does not make adequate allowance for variations in fuel, air, and exhaust gas compositions through the fuel cell. As a consequence of this, there is a choice of whether to select the inlet, outlet or an average value of the gas composition in the fuel cell global equations, with each selection achieving significantly different results.

Magistri *et al.* [4] compared simplified with detailed SOFC models, and tested how the level of detail affected the predictions of the design-point performance of hybrid systems. They emphasized the usefulness of the simplified model for hybrid system design and off-design analysis, and the requirement of a detailed model to generate a more complete description of the SOFC internal behaviour.

Judkoff and Neymark [5] reported and classified into three groups the sources of simulation errors encountered when using building simulation programs (these errors are applicable to SOFC HS simulation):

1) errors introduced due to assumptions and simplifications;
2) errors or inaccuracies in solving mathematical equations;
3) coding errors.

They also proposed a pragmatic, three-step approach to identifying these errors. The first approach consists of comparative testing: the results of the model should be compared with the results of other models of benchmark quality for the same problem with similar initial and boundary conditions. If the results of the models match to within an acceptable tolerance, this implies the particular implementations are acceptable. However, meeting this validation target does not guarantee that an implementation will be correct in different applications or for different problems. The models can still contain significant errors or have limitations. In the second approach, based on analytical validation procedures, the results of the model for a simple case are compared with the results of an available analytical solution (if this exists). Finally, in the empirical validation approach, the results of the simulations are compared with real data from the actual system under laboratory or field conditions [1].

The validation of a model is necessary because a calculation tool must be compared with reference cases to be considered able to produce reliable results. Appropriate data are needed for validation. With limited resources, this can be difficult as most data are not freely available in the open literature. Even though performance data from an entire hybrid power generation system are usually proprietary and not available in the literature, information for a single component is easier to find. Consequently, a way to resolve the problem of limited performance data is to develop and validate well-defined subsystem models, and then integrate them together into a complete model of a large hybrid power generation system [1].

Even though the SOFC is considered to be the 'heart' of these hybrid cycles, its detailed mathematical modelling and simulation methodology is not included in this review. The focus here is on the evaluation of the overall system performance and not on its component performance. Further details and reviews of SOFC modelling can be found in Zabihian and Fung [1] and references therein. In addition, some good examples of such simulations can be found in [6, 7] for steady-state and [8, 9] for transient and dynamic modelling.

With respect to SOFC hybrid systems, numerous computational models have been developed by various authors and presented in the literature. These models have been used to analyse SOFC HS performance, including off-design behaviour at component and system levels. For component calculations, detailed approaches are employed. This is particularly true for the fuel cell, for which access may be required to a large number of design and development variables, especially when considering, for example, loss mechanisms, enabling technologies, and optimization procedures. SOFC subcomponents also often require particularly detailed attention. For instance, material-related aspects of electrode design as well as electrolyte selection and performance issues require very complex chemical models involving micro-kinetic descriptions. Studies have been presented in the literature employing complex 3D CFD tools incorporating these detailed electrochemical reactions to describe the relevant fluid dynamic and transport aspects of the design, involving relevant variable/property field distributions for SOFCs [10–12]. If real-time performance is required, however, modelling approaches have to be extremely simplified while capturing most of the engineering considerations of interest [13]; interpolation between performance maps is usually necessary when the physical aspects are too complicated.

5.3.1 Computational Models for Hybrid Systems at the Component Level

This section presents some simple guidance, describes procedures and gives practical examples for hybrid system model development at the component level using a simple 0D approach involving control volume analyses. Since these systems include a microturbine (in some cases based on complex layouts, such as in a multishaft machine), the modelling discussion starts with the presentation of the fundamental techniques for calculating the non-equilibrium off-design performance of gas turbines (from simple to more complex layouts).

Matching conditions in SOFC hybrid systems are different from what is necessary to calculate for a standard machine, because fuel cells generate greater pressure drops and produce flows with a different chemical composition and different temperature values (also the turbine inlet temperature is due to SOFC operative performance, because it does not work with the flexibility of a standard combustor) at the turbine inlet. The most commonly considered gas turbines for hybrid systems are based on the simple cycle, designed with a single shaft turbine, but a number of other configurations are possible. For this reason, a general discussion on turbine off-design is presented.

5.3.2 Prediction of Performance of Gas Turbines

From cycle calculations it is possible to design the individual components of a gas turbine so that the complete unit will give the required performance when running at the *design point*, that is, when it is running at a defined speed, pressure ratio and mass flow for which the components are designed. However, it is also necessary to find the performance variation of the gas turbine over the complete operating range of speed and power output, which is normally referred to as the *off-design* performance [14]. This process involves finding corresponding operating points for the characteristics of each component when the machine is running at steady-state mode under different conditions from the design point. These equilibrium running points for a series of rotational speeds may then be plotted on the compressor characteristic map and joined up to form an *equilibrium running line* forming an *equilibrium running diagram*. Once the operating conditions have been determined, it is relatively simple to obtain performance curves of power output or thrust, and specific fuel consumption [14]. The following points below summarize the characteristics of the equilibrium diagram. These aspects are also necessary to remember when predicting the performance of gas turbines.

1) The equilibrium running diagram shows the proximity of the operating line or zone to the compressor surge line.
2) The variation of specific fuel consumption with reduction in power, sometimes referred to as *part-load* performance, is of major importance in applications where considerable running at low power settings is required.
3) When determining the off-design performance it is important to be able to predict not only the effect of operation at part-load on specific fuel consumption, but also the effect of ambient conditions on maximum output.

The types of gas turbine discussed in this chapter will include (Figure 5.2):

- the single-shaft unit;
- the free turbine engine, where the gas-generator turbine drives the compressor and the power turbine drives the load.

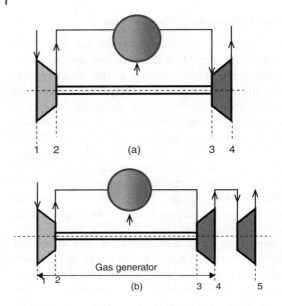

Figure 5.2 (a) Simple gas turbine layout; and (b) double-shaft gas turbine scheme.

Figure 5.3(a) shows the compressor characteristic maps. With high-performance axial compressors, the constant speed lines become vertical on a mass flow basis when the inlet is choked, while for radial machines (such as microturbines) the trend is more flat. Due to the choking behaviour, the mass flow function of turbines can be represented by a single curved line as seen in Figure 5.3(b).

For accurate calculations it is necessary to consider the variation of pressure losses in the inlet duct, the combustion chamber and the exhaust duct. For the design or selection of a turbine for use in a hybrid system, it is necessary to determine the right matching conditions.

5.3.3 Off-design Operation of the Single-shaft Gas Turbine

This is the procedure for obtaining an equilibrium running point:

1) Select a constant speed line on the compressor characteristic and choose any point on this line; determine the values of $m\sqrt{T_{01}}/p_{01}$, $p_{02}/p_{01}\eta_c$ and $N/\sqrt{T_{01}}$.

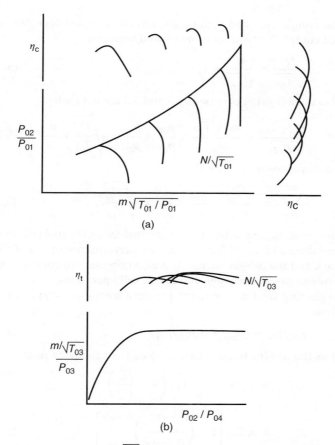

(a)

(b)

Note: parameters will be $\dfrac{m\sqrt{T_{04}}}{P_{04}} \cdot \dfrac{P_{04}}{P_2} \cdot \dfrac{N_p}{\sqrt{T_{04}}}$ and η_{tp} for a power turbine

Figure 5.3 Compressor and turbine characteristics. *Source*: Cohen *et al.*
(1996) [14]. Reproduced with permission from Pearson.

2) The corresponding point on the turbine characteristic is
   obtained from consideration of compatibility of rotational
   speed and flow.
3) Having matched the compressor and turbine characteristics,
   it is necessary to ascertain whether the work output corresponding to the selected operating point is compatible with
   that required by the driven loads.

The compressor and turbine are directly coupled together, so that compatibility of rotational speed requires:

$$\frac{N}{\sqrt{T_{03}}} = \frac{N}{\sqrt{T_{01}}} \times \sqrt{\frac{T_{01}}{T_{03}}} \tag{5.1}$$

Also the following equations 5.2 and 5.3 are necessary.

$$\frac{m_3\sqrt{T_{03}}}{p_{03}} = \frac{m_1\sqrt{T_{01}}}{p_{01}} \times \frac{p_{01}}{p_{02}} \times \frac{p_{02}}{p_{03}} \times \sqrt{\frac{T_{03}}{T_{01}}} \times \frac{m_3}{m_1} \tag{5.2}$$

Assuming $m_1 = m_3 = m$ Then

$$\frac{\sqrt{T_{03}}}{p_{03}} = \frac{\sqrt{T_{01}}}{p_{01}} \times \frac{p_{01}}{p_{02}} \times \frac{p_{02}}{p_{03}} \times \sqrt{\frac{T_{03}}{T_{01}}} \tag{5.3}$$

Moreover, p_{03} has to be calculated considering the fuel cell pressure drop and T_{03} is linked with the performance of the SOFC stack and the off-gas burner. It is also necessary to consider the flow composition change due to fuel cell operations.

Neglecting inlet and exhaust pressure losses, $p_a = p_{01} = p_{04}$. Then:

$$p_{03}/p_{04} = (p_{03}/p_{02})(p_{02}/p_{01}) \tag{5.4}$$

Thus the turbine temperature drop can be calculated from:

$$\Delta T_{034} = T_{03} - T_{04} = T_{03}\left(1 - \frac{T_{04}}{T_{03}}\right)$$

$$= \eta_t T_{03}\left[1 - \left(\frac{1}{p_{03}/p_{04}}\right)^{(\gamma-1)/\gamma}\right] \tag{5.5}$$

The compressor temperature rise can be similarly calculated by:

$$\Delta T_{012} = \frac{T_{01}}{\eta_c}\left[\left(\frac{p_{02}}{p_{01}}\right)^{(\gamma-1)/\gamma} - 1\right] \tag{5.6}$$

Net power output is equal to:

$$mc_{pg}\Delta T_{034} - \frac{1}{\eta_m}mc_{pa}\Delta T_{012} \tag{5.7}$$

where η_m is the mechanical efficiency of the compressor-turbine combination:

$$(m\sqrt{T_{03}}/p_{01})(p_a/\sqrt{T_a}) \tag{5.8}$$

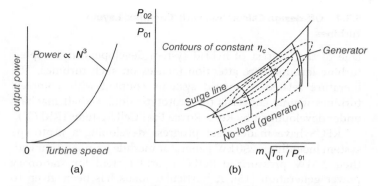

Figure 5.4 (a) Load characteristics and (b) equilibrium running lines. *Source*: Cohen *et al*. (1996) [14]. Reproduced with permission from Pearson.

Finally, it is necessary to consider the characteristics of the load to determine whether the compressor operating point selected represents a valid solution. With an electrical generator configuration for a load, for example, the power absorbed varies as the cube of the rotational speed of the machine. When the transmission efficiency and gear ratio are known, the load characteristic in terms of the net power actually required from the turbine and the turbine speed can be plotted as in Figure 5.4(a). Then it is necessary to find the single point on each constant speed line of the compressor characteristic that will give the required net power output at that speed.

If the calculated net power output for any point on the compressor characteristic is not equal to the power required at the selected speed, the engine will not be in equilibrium and will either accelerate or decelerate depending on whether there is an excess or deficiency of power. Repeating this procedure for a series of constant speed lines, a series of points is obtained to form the equilibrium running line.

The equilibrium running lines depicted in Figure 5.4(b) show that the example implies operation in a zone of high compressor efficiency over a wide range of output. The location of the equilibrium running line relative to the surge line indicates whether the engine can be brought up to full power without any complications.

5.3.4 Off-design Calculation with 'Complex' Layout Turbines

Due to the interest of hybrid system developers in 'complex' turbine layouts, here attention focuses on such turbines. The literature related to hybrid systems coupled with 'complex' turbines mainly concerns the prototype double-shaft machine under development by Rolls-Royce Fuel Cell Systems (RRFCS).

RRFCS have made great progress developing a prototype system including a 250 kW generator module for operation with their 1 MW pressurized SOFC hybrid system for stationary power generation [15]. A particular focus has been given to the system efficiency and the improvement obtained with pressurization. A specially designed turbo generator, operating with two shafts [15], provides the pressure increase necessary in the RRFCS hybrid system cathodic ejector.

Cohen *et al.* [14], in Chapter 5 of their book, have given some simple examples of gas turbine performance prediction involving the matching of a gas generator with a free power turbine. The engineering procedure is based on a mass-matching algorithm considering the mass flow leaving the gas generator to be compatible with that at entry to the power turbine. This is coupled with the fact that the pressure ratio available to the power turbine is fixed by the compressor and gas-generator turbine pressure ratios. They also discuss the matching of two turbines in series and indicate that the iterative procedure required for the matching of a gas generator and a free turbine can be simplified if reduced-order running of the two turbines in series can be assumed; this approach is also valuable for the analysis of more complex gas turbines, as considered in Chapter 9 of [14].

5.3.4.1 Equilibrium Running of a Gas Generator

Under equilibrium conditions the pressure ratio across the turbine can be determined by equating the turbine work to the compressor work. The required turbine temperature drop, in conjunction with the turbine inlet temperature and efficiency, determines the turbine pressure ratio by:

$$\eta_m c_{pg} \Delta T_{034} = c_{pa} \Delta T_{012} \tag{5.9}$$

Rearranging Equation 5.9, it is possible to obtain:

$$\frac{\Delta T_{034}}{T_{03}} = \frac{\Delta T_{012}}{T_{01}} \times \frac{T_{01}}{T_{03}} \times \frac{c_{pa}}{c_{pg}\eta_m} \tag{5.10}$$

5.3.4.2 Off-design Operation of a Free Turbine Engine

The gas generator is matched to the power turbine through the mass flow rate. The mass flow leaving the gas generator must be equal to the mass flow entering the power turbine. In addition, the pressure ratio available to the power turbine is fixed by the compressor and gas-generator turbine pressure ratio. The preceding section described how the gas-generator operating conditions can be determined for any point on the compressor characteristic. Now the matching equations for the free turbine are presented (Equations 5.11–5.13).

$$\frac{m_3\sqrt{T_{04}}}{p_{04}} = \frac{m_1\sqrt{T_{03}}}{p_{03}} \times \frac{p_{03}}{p_{04}} \times \sqrt{\frac{T_{04}}{T_{03}}} \tag{5.11}$$

Then

$$\sqrt{\frac{T_{04}}{T_{03}}} = \sqrt{1 - \frac{\Delta T_{034}}{T_{03}}}$$

where

$$\frac{\Delta T_{034}}{T_{03}} = \eta_t \left[1 - \left(\frac{1}{p_{03}/p_{04}} \right)^{(\gamma-1)/\gamma} \right] \tag{5.12}$$

and

$$\frac{p_{04}}{p_a} = \frac{p_{02}}{p_{01}} \times \frac{p_{03}}{p_{02}} \times \frac{p_{04}}{p_{03}} \tag{5.13}$$

Having found the pressure ratio across the power turbine, the value of $\frac{m_3\sqrt{T_{04}}}{p_{04}}$ can be found from the power turbine characteristic for comparison with the value obtained from the above equation. If agreement is not reached, it is necessary to choose another point on the same constant speed line of the compressor characteristic and repeat the procedure until the requirement of flow compatibility between the two turbines is satisfied. For each constant speed line on the compressor characteristic, there will be only one point that will satisfy both the work requirement

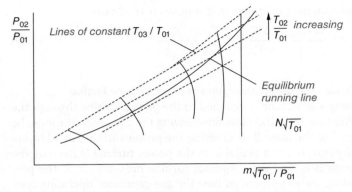

Figure 5.5 Equilibrium running line for a free turbine. *Source*: Cohen *et al.* (1996) [14]. Reproduced with permission from Pearson.

of the gas generator and the flow compatibility with the power turbine. The equilibrium running line can be obtained if foregoing calculations are carried out for each speed line. The running line for the free turbine engine is independent of the load and is determined by the single-shaft capacity of the power turbine, as shown in Figure 5.5.

The iterative procedure required to match a gas generator and a free turbine can be simplified if the behaviour of two turbines in series is considered.

In practice, the variation of turbine efficiency at any given pressure ratio is not large, particularly over the restricted range of operation of the gas-generator turbine. It is often sufficiently accurate to take a mean value of turbine efficiency at any given pressure ratio, so that the reduced mass flow rate for point 4 becomes a function only of p_{03}/p_{04} and the reduced mass flow rate for point 3 (dotted curve in Figure 5.6).

In particular, as long as the power turbine is choked, the gas-generator turbine will operate at a fixed non-dimensional point, that is at a fixed pressure ratio marked (a) in Figure 5.6. With the power turbine unchoked, the gas-generator will be restrained to operate at a fixed pressure ratio for each power turbine ratio, that is, (b) and (c) in Figure 5.6. Thus the maximum pressure ratio across the gas-generator turbine is controlled by choking of the power turbine.

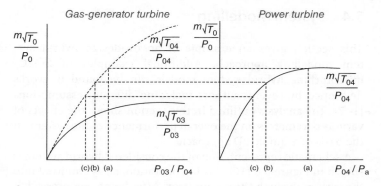

Figure 5.6 Equilibrium running line for free turbine. *Source*: Cohen *et al.* (1996) [14]. Reproduced with permission from Pearson.

A further consequence of the fixed relationship between the turbine pressure ratios is:

$$\frac{P_{03}}{P_{04}} = \frac{P_{02}}{P_{01}} \times \frac{P_{03}}{P_{02}} \times \frac{P_a}{P_{04}} \tag{5.14}$$

$\frac{P_{03}}{P_{02}}$ is determined by the assumed combustion pressure loss, and $\frac{P_a}{P_{04}}$ is obtained from the above figure.

Thus for each constant speed line considered, only a single iteration is required to find the correct equilibrium running point. Such a curve is shown in Figure 5.7.

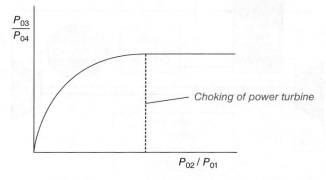

Figure 5.7 Equilibrium running line for free turbine. *Source*: Cohen *et al.* (1996) [14]. Reproduced with permission from Pearson.

5.4   System Modelling

This section gives an example of an on-design hybrid system modelling approach. Additional examples of SOFC-microturbine hybrid system models can be found in works developed by several authors including different assumptions [1–7]. The analysis outlined in this section shows the effects of various parameters on the overall performance of the system on the basis of a simple 0D approach.

A schematic representation of the considered hybrid system is shown in Figure 5.8 [16]. During operation, air is admitted into the system through the compressor. After being compressed, it is fed into a recuperator (REC) where the waste heat from the turbine exhaust is recovered to increase the SOFC inlet temperature. The heated air then goes into the cathode side of the fuel cell, where oxygen molecules are consumed in the electrochemical reactions. Air in this example is assumed to be composed of 21% oxygen and 79% nitrogen by volume respectively.

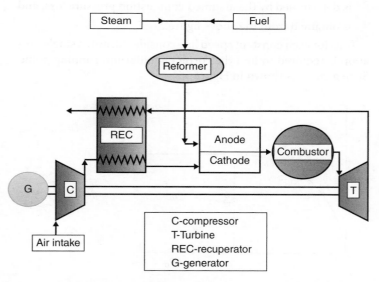

Figure 5.8 Diagram of the hybrid system considered for this modelling activity.

For the fuel stream, this example assumes a composition of 96% methane, 2% nitrogen and 2% carbon dioxide by volume. Fuel supply is first mixed with the steam according to a designed steam to carbon ratio (S/C) to prevent carbon deposition. Anodic recirculation technology [4, 13], even if optimal for such system operations, is not included in this example for simplicity. The mixture then enters the reformer (an internal reforming unit) where fuel reforming occurs. The required endothermic heat for reforming is assumed to be provided from the residual heat within the SOFCs, as the SOFC (see Chapter 1) generates a substantial amount of waste heat much greater than required by the reformer [17]. After the reformer, a hydrogen-rich mixture enters the SOFC for electrochemical conversion.

Exhausts from the anode and cathode are then mixed into an additional combustor unit before admission into the turbine. In the combustor, the remaining hydrogen, carbon monoxide and methane are burnt completely to increase the temperature of the exhaust. Use of this device avoids the emission of unused fuel in the exhaust gas and allows for a higher amount of work to be produced by the turbine. The exhaust from the turbine is diverted to the recuperator to transfer energy to the compressed air, recovering as much energy as possible.

To simplify the case study, the exhaust from the reformer is assumed to be of the same temperature as the stream in the reformer. The same assumption also applies to the SOFC [18–21].

Before proceeding with the modelling of individual components, some generic assumptions have been made.

1) All components in the system are adiabatic.
2) Temperature, gas composition and pressure in each component are uniformly distributed.
3) Chemical reactions are at equilibrium.
4) There are no leaks in the system.

5.4.1 Reformer

According to the S/C, a certain amount of steam is required by the reforming process. This steam can either be generated using waste heat from the fuel cell or from the turbine exhaust

Table 5.1 Values of constants used in reforming and water–gas shifting process.

	Reforming	Shifting
a_1	-2.63121×10^{-11}	5.47301×10^{-12}
a_2	1.24065×10^{-7}	-2.57479×10^{-8}
a_3	-2.25232×10^{-4}	4.63742×10^{-5}
a_4	0.195028	-0.03915
a_5	-66.1395	13.2097

Source: Adapted from Nien (2007) [16].

after the recuperator. Two reactions previously mentioned (equations 5.15 and 5.16) take place in the reformer.

$$CH_4 + H_2O \rightarrow CO + 3H_2 \tag{5.15}$$

$$CO + H_2O \rightarrow CO_2 + H_2 \tag{5.16}$$

To analyze the reactions, their equilibrium constants have to be evaluated. These constants are temperature-dependent, and polynomial correlations (Equation 5.17) have been developed to evaluate them [22–24].

$$\log K_p = a_1 T^4 + a_2 T^3 + a_3 T^2 + a_4 T + a_5 \tag{5.17}$$

where a_1, a_2, a_3, a_4 and a_5 are constants listed in Table 5.1.

With reforming and shifting reactions always in equilibrium, the equilibrium constants can also be obtained from partial pressures of the products. Referring to Equations 5.15–5.16, it can be observed that equilibrium constants can be derived from the two expressions below, where K_{pr} and K_{ps} are reforming and shifting equilibrium constants respectively.

$$K_{pr} = \left(\frac{P_{H_2}{}^3 P_{CO}}{P_{H_2O} P_{CH_4}} \right) \tag{5.18}$$

$$K_{ps} = \left(\frac{P_{H_2} P_{CO_2}}{P_{H_2O} P_{CO}} \right) \tag{5.19}$$

If the percentage flow rates of CH_4 and CO are assumed to be x and y respectively, with the initial flow of H_2 being A, the outlet molar flow of H_2 can be represented by the expression

Table 5.2 Inlet and outlet percentage flows of the reformer.

Component	Inlet flow percentage	Outlet flow percentage
H_2	A	$A + 3x + y - z$
H_2O	B	$B - x - y + z$
CH_4	C	$C - x$
CO	D	$D + x - y$
CO_2	E	$E + y$
Total	$A + B + C + D + E = 1$	$A + B + C + D + E + 2x = 1 + 2x$

Source: Adapted from Nien (2007) [16].

$A + 3x + y$. Table 5.2 shows the percentage flow rates of all components before and after reformer.

As the partial pressure of a component is the product of its molar percentage and total pressure, the partial pressures of components can be expressed by these two quantities. If expressions for outlet molar flows are substituted into Equation 5.18, the equation then becomes:

$$K_{pr} = \left(\frac{\left(\frac{A+3x+y}{1+2x}Pt\right)^3 \left(\frac{D+x-y}{1+2x}Pt\right)}{\left(\frac{B-3x-y}{1+2x}Pt\right)\left(\frac{C-x}{1+2x}Pt\right)} \right)$$

$$K_{pr} = \left(\frac{\left(\frac{A+3x+y}{1+2x}\right)^3 \left(\frac{D+x-y}{1+2x}\right)}{\left(\frac{B-x-y}{1+2x}\right)\left(\frac{C-x}{1+2x}\right)} \right) Pt^2 \qquad (5.20)$$

where Pt is the pressure in the reformer. Notice that the percentage flow used here as the initial molar flow is unknown.

Similarly for Equation 5.19, it is possible to obtain:

$$K_{ps} = \left(\frac{\left(\frac{A+3x+y}{1+2x}Pt\right) \left(\frac{E+y}{1+2x}Pt\right)}{\left(\frac{B-x-y}{1+2x}Pt\right)\left(\frac{D+x-y}{1+2x}Pt\right)} \right)$$

$$K_{ps} = \left(\frac{\left(\frac{A+3x+y}{1+2x}\right) \left(\frac{E+y}{1+2x}\right)}{\left(\frac{B-x-y}{1+2x}\right)\left(\frac{D+x-y}{1+2x}\right)} \right)$$

$$K_{ps} = \left(\frac{(A + 3x + y)(E + y)}{(B - x - y)(D + x - y)} \right) \tag{5.21}$$

If x is assumed as an initial value and the relevant K_ps are known from tables or derived elsewhere from fundamental considerations, y can be calculated. Equation 5.21 can be modified to calculate the value of y as follows:

$$K_{ps}(B - x - y)(D + x - y) = (A + 3x + y)(E + y) \tag{5.22}$$

$$K_{ps}\left(BD + Bx - By - Dx - x^2 + xy - Dy - xy + y^2 \right)$$
$$= AE + Ay + 3Ex + 3xy + Ey + y^2 \tag{5.23}$$

$$y^2 \left(K_{ps} - 1 \right) + y \left(-K_{ps}B - K_{ps}D - E - A - 3x \right)$$
$$+ \left(K_{ps}BD + K_{ps}Bx - K_{ps}Dx - K_{ps}x^2 - AE - 3Ex \right) = 0 \tag{5.24}$$

Since the inlet flow to the reformer does not contain CO and H_2, $A = 0$ and $D = 0$. Therefore the above equation can be simplified to:

$$y^2 \left(K_{ps} - 1 \right) + y \left(-K_{ps}B - E - 3x \right)$$
$$+ \left(K_{ps}Bx - K_{ps}x^2 - 3Ex \right) = 0 \tag{5.25}$$

Comparing this equation with

$$ax^2 + bx + c = 0 \tag{5.26}$$

where

$$a = K_{ps} - 1$$
$$b = -K_{ps}B - E - 3x$$
$$c = K_{ps}Bx - K_{ps}x^2 - 3Ex$$

With a, b and c evaluated, y can be obtained using the expression below:

$$y = \frac{-b \pm \sqrt{b^2 - 4ac}}{2a} \tag{5.27}$$

Now, after assuming a reasonable value of x and calculating a value of y, K_{pr} can be calculated. Iterations are performed until the calculated value of K_{pr} is equal to the prior K_{pr} obtained from Equation 5.16. Subsequently, the fuel composition after the reformer can be determined.

5.4.2 SOFC Module

The initial overriding parameter to be calculated for the fuel cell is electrochemical voltage produced by the cell. The fuel cell used in this study is a tubular SOFC. In the introductory section of Chapter 1, a detailed discussion regarding the nature of the Nernst's voltage (reversible cell voltage at a certain pressure and temperature) was provided. It was emphasized that the cell voltage is significantly affected by the prevailing cell temperature, pressure and concentration in an expression given by the Nernst's equation (5.28).

$$E = E^0 + \frac{RT}{2F} \ln \left(\frac{\alpha \beta^2}{\gamma} \right) + \frac{RT}{4F} \ln(P) \qquad (5.28)$$

The second term in the equation can be determined using the fuel composition calculated in the reformer model. The only unknown in the third term is pressure, which is easily obtainable. E^0 is the reversible cell voltage at standard atmospheric pressure and can be calculated by:

$$E = -\frac{\Delta \bar{g}_f}{2F} \qquad (5.29)$$

The change in molar Gibbs free energy has to be initially calculated. Equation 5.30 shows that molar Gibbs free energy comprises contributions from the molar enthalpy and entropy change.

$$\Delta \bar{g}_f = \Delta \bar{h}_f - T \Delta \bar{S} \qquad (5.30)$$

Considering the electrochemical reaction (5.31) it is possible to obtain Equation 5.32.

$$H_2 + \tfrac{1}{2}O_2 \rightarrow H_2O \qquad (5.31)$$

$$\Delta \bar{h}_f = \Delta \bar{h}_{f \cdot products} - \Delta \bar{h}_{f \cdot reactants}$$

$$\Delta \bar{h}_f = \bar{h}_{f \cdot H_2O} - \bar{h}_{f \cdot H_2} - \tfrac{1}{2}\bar{h}_{f \cdot O_2} \qquad (5.32)$$

Similarly, it is possible to obtain Equation 5.33 for the entropy.

$$\Delta \bar{s} = \bar{s}_{H_2O} - \bar{s}_{H_2} - \tfrac{1}{2}\bar{s}_{O_2} \qquad (5.33)$$

Since the specific electrochemical reaction takes place in SOFCs, the enthalpy of formation and entropy are evaluated

Table 5.3 Coefficients in equations to evaluate enthalpy, entropy and specific heat capacity.

Species	a_1	a_2	a_3	a_4	a_5	a_6	a_7
300 K < T < 1000 K							
CH_4	1.503	1.042×10^{-2}	-3.918×10^{-6}	6.778×10^{-10}	-4.428×10^{-14}	-9.979×10^3	10.71
CO_2	4.461	-3.098×10^{-3}	-1.239×10^{-6}	2.274×10^{-10}	-1.553×10^{-14}	-4.896×10^4	−0.9864
H_2	3.100	5.112×10^{-4}	5.264×10^{-8}	-3.491×10^{-11}	3.695×10^{-15}	-8.774×10^2	−1.963
H_2O (G)	2.717	2.945×10^{-3}	-8.022×10^{-7}	1.023×10^{-10}	-4.847×10^{-15}	-2.991×10^4	6.631
H_2O (L)	12.71	-1.766×10^{-2}	-2.256×10^{-5}	2.082×10^{-7}	-2.408×10^{-10}	-3.748×10^4	−59.12
N_2	2.896	1.515×10^{-3}	-5.724×10^{-7}	9.981×10^{-11}	-6.522×10^{-15}	-9.059×10^2	6.162
O_2	3.622	7.362×10^{-4}	-1.965×10^{-7}	3.620×10^{-11}	-2.895×10^{-15}	-1.202×10^3	3.615
CO	3.580	-6.104×10^{-4}	1.017×10^{-6}	9.070×10^{-10}	-9.044×10^{-13}	-1.434×10^4	3.508
1000 K < T < 5000 K							
CH_4	3.826	-3.979×10^{-3}	2.456×10^{-5}	-2.273×10^{-8}	6.963×10^{-12}	-1.014×10^4	0.8669
CO_2	2.401	8.735×10^{-3}	-6.607×10^{-6}	2.002×10^{-9}	6.327×10^{-16}	-4.838×10^4	9.695
H_2	3.057	2.677×10^{-3}	-5.810×10^{-6}	5.521×10^{-9}	-1.812×10^{-12}	-9.889×10^2	−2.300
H_2O (G)	4.070	-1.108×10^{-3}	4.152×10^{-6}	-2.964×10^{-9}	8.070×10^{-13}	-3.028×10^4	−0.3227
H_2O (L)	12.71	-1.766×10^{-2}	-2.256×10^{-5}	2.082×10^{-7}	-2.408×10^{-10}	-3.748×10^4	−59.12
N_2	3.675	-1.208×10^{-3}	2.324×10^{-6}	-6.322×10^{-10}	-2.258×10^{-13}	-1.061×10^3	2.358
O_2	3.626	-1.878×10^{-3}	7.055×10^{-6}	-6.764×10^{-9}	2.156×10^{-12}	-1.048×10^3	4.305
CO	2.715	2.063×10^{-3}	-9.988×10^{-7}	2.301×10^{-10}	-2.036×10^{-14}	-1.415×10^4	7.819

Source: Adapted from Nien (2007) [16].

at the SOFC temperature, using polynomial equations fitted via data tables provided by the Joint Army-Navy-Air Force (JANAF) project [25–28]:

$$\frac{h_f}{RT} = a_1 + \frac{a_2}{2}T + \frac{a_3}{3}T^2 + \frac{a_4}{4}T^3 + \frac{a_5}{5}T^4 + \frac{a_6}{6}T \quad (5.34)$$

$$\frac{s}{R} = a_1 \ln T + a_2 T + \frac{a_3}{3}T^2 + \frac{a_4}{4}T^3 + \frac{a_5}{5}T^4 + a_7 \quad (5.35)$$

where a_1 to a_7 are listed in Table 5.3 with enthalpy of formation and entropy for individual gases. The reversible cell voltage at standard pressure can be found using equations 5.32, 5.33, 5.29 and 5.30.

5.4.3 Overpotentials

The next step is to obtain the sum of overpotentials (V_{loss}) so that the net operating voltage of the fuel cells can be determined:

$$V_{OP} = E - V_{loss} \quad (5.36)$$

V_{loss} in this study is modelled by three losses: overpotential at the anode (η_a), overpotential at the cathode (η_c) and ohmic losses (η_{ohmic}) [29]. Equations in 5.37 are used to evaluate these losses for SOFC temperature at 1273 K [29, 30]. Details of SOFC dimensions can be found in Table 5.4. For more details on these empirical equations, the reader is referred to [29, 30].

Table 5.4 SOFC parameters for calculation of ohmic losses.

Interconnector		Cathode	
Resistivity (Ω cm), σic	0.5	Resistivity (Ω cm), σc	0.015
Thickness (cm), dic	0.002	Thickness (cm), dc	0.2
Width (cm), wic	0.5		
Anode		**Electrolyte**	
Resistivity (Ω cm), σa	0.001	Resistivity (Ω cm), σe	10
Thickness (cm), da	0.01	Thickness (cm), de	0.002
		Cell diameter (cm), I	1.6

Source: Adapted from Nien (2007) [16] and Tanaka *et al.* (2000) [29].

$$\left.\begin{array}{l} \eta_a = 0.05 \\ \eta_c = 0.07j^3 - 0.077j^2 + 0.04j - 0.00003 \\ \eta_{ohmic} = j\sigma_e d_e + jl\pi_{ic}\frac{d_{ic}}{w_{ic}} + \frac{jl^2\pi^2}{8}\left(\frac{\sigma_c}{d_c} + \frac{\sigma_a}{d_a}\right) \end{array}\right\} \qquad (5.37)$$

5.4.4 Fuel and Air Supply Calculations

Considering the electrochemical reaction in the fuel cell allows for the calculation of the required fuel and air flows [31].

$$\text{Cathode:} \quad O_2 + 4e^- \rightarrow 2O^{2-} \qquad (5.38)$$

$$\text{Anode:} \quad 2H_2 + 2O^{2-} \rightarrow 2H_2O + 4e^- \qquad (5.39)$$

With reference to reaction 5.39, it is clear that for every two moles of electrons created, one mole of hydrogen is reacted. Therefore, I (molar flow rate of electrons) can be calculated considering equations 5.40 and 5.41:

$$I = \left(\dot{n}_{H_2}\right)_{reacted}\cdot 2F \qquad (5.40)$$

Since $P_{sofc} = VI$,

$$P_{sofc} = V\cdot\left(\dot{n}_{H_2}\right)_{reacted}\cdot 2F \qquad (5.41)$$

Rearranging the above equation it is possible to obtain:

$$\left(\dot{n}_{H_2}\right)_{reacted} = \frac{P_{sofc}}{2VF} \quad \text{(moles/s)} \qquad (5.42)$$

When a similar analysis is performed on oxygen using Equation 5.38, we obtain:

$$\left(\dot{n}_{O_2}\right)_{reacted} = \frac{P_{sofc}}{4VF} \quad \text{(moles/s)} \qquad (5.43)$$

Since the electrochemical reaction does not consume all the hydrogen and oxygen in the flow stream [28], they are usually supplied in excess amounts. Fuel and air utilization factors govern the relationships between the amount consumed and the amount supplied for hydrogen and oxygen respectively.

In the case of a cell working with hydrogen as the only fuel, the fuel utilization factor, U_f, is defined as the ratio between the amount of hydrogen reacted and the amount supplied in case of a cell working with hydrogen as only fuel:

$$U_f = \frac{\left(\dot{n}_{H_2}\right)_{reacted}}{\left(\dot{n}_{H_2}\right)_{supplied}} \qquad (5.44)$$

Rearranging this equation gives:

$$\left(\dot{n}_{H_2}\right)_{supplied} = \frac{\left(\dot{n}_{H_2}\right)_{reacted}}{U_f} \quad (moles/s) \tag{5.45}$$

The air utilization factor (U_A) is the ratio between the reacted and supplied oxygen. Equation 5.47 is used to calculate the oxygen flow to be supplied.

$$U_A = \frac{\left(\dot{n}_{O_2}\right)_{reacted}}{\left(\dot{n}_{O_2}\right)_{supplied}} \tag{5.46}$$

$$\left(\dot{n}_{O_2}\right)_{supplied} = \frac{\left(\dot{n}_{O_2}\right)_{reacted}}{U_A} \quad (moles/s) \tag{5.47}$$

Because the supply stream is made up of 21% oxygen, the total molar flow rate of air can be calculated using:

$$\left(\dot{n}_{air}\right)_{supplied} = \frac{\left(\dot{n}_{O_2}\right)_{supplied}}{0.21} \quad (moles/s) \tag{5.48}$$

5.4.5 Combustor

The inlet flows to the combustor are the exhausts from the anode and cathode outlet ducts. To determine combustor outlet temperature and conditions, a combustion efficiency of 100% is assumed (as most combustors have efficiencies in the high 90s). Therefore the mathematical model assumes the remaining CH_4, CO, and H_2 from the SOFC exhaust are burned completely [32]. The following three reactions thus proceed to completion:

$$CH_4 + 2O_2 \rightarrow CO_2 + 2H_2O \tag{5.49}$$
$$CO + \tfrac{1}{2}O_2 \rightarrow CO_2 \tag{5.50}$$
$$H_2 + \tfrac{1}{2}O_2 \rightarrow H_2O \tag{5.51}$$

From the prior assumptions, combustor outlet molar concentrations can be found: $CH_{4out} = 0$; $CO_{out} = 0$; $H_{2out} = 0$; $CO_{2out} = CO_{2in} + CO_{in} + CH_{4in}$; $O_{2out} = O_{2in} - \tfrac{1}{2}CO_{in} - 2CH_{4in} - \tfrac{1}{2}H_{2in}$; $H_2O_{out} = H_2O_{in} + H_{2in} + 2CH_{4in}$, and $N_{2out} = N_{2in}$.

The technique to determine the outlet temperature is based on an iterative procedure: assuming an outlet temperature (inlet temperature is used for this assumption) and carrying out an enthalpy calculation for the products at the assumed temperature. The calculation is then repeated until a satisfactory convergence is achieved.

5.4.6 Turbine

The turbine is assumed to be adiabatic, and hence the power generated by the turbine is given by [24, 32]:

$$W_t = m \cdot c_p \cdot \eta_t \cdot T_i \cdot \left(\left(\frac{P_O}{P_I} \right)^{\frac{\gamma-1}{\gamma}} - 1 \right) \qquad (5.52)$$

where m is the mass flow rate (kg/s), c_p the specific heat (J kg^{-1} K^{-1}), η_t is the turbine adiabatic efficiency, T_i is the inlet temperature (K), $\frac{P_O}{P_I}$ is the turbine expansion ratio and γ is the ratio of specific heat of the gas, $\frac{c_p}{c_v}$.

In Equation 5.52, the mass flow rate, inlet temperature, inlet pressure and turbine efficiency are known parameters. P_O can be easily determined by assuming the outlet pressure of the turbine is equal to the atmospheric value plus the known recuperator pressure loss value. Since the inlet flow composition and flow rates have been determined in the combustor model, it is possible to calculate c_p for the flow. c_p and c_v are the molar specific heats at constant pressure and volume respectively, while R (8.3144 kJ/kmol K) is the universal gas constant.

$$c_p - c_v = R$$
$$\frac{c_p}{c_v} = \gamma$$
$$\frac{\gamma - 1}{\gamma} = \frac{\frac{c_p}{c_v} - 1}{\frac{c_p}{c_v}} = \frac{c_p - c_v}{c_p} = \frac{R}{c_p} \qquad (5.53)$$

Therefore, if c_p is known, all the remaining unknown values can be calculated in Equation 5.52. c_p of individual species can be calculated using this polynomial expression [26, 27]:

$$\frac{c_p}{R} = a_1 + a_2 T + a_3 T^2 + a_4 T^3 + a_5 T^4 \qquad (5.54)$$

5.4.7 Compressor

The compressor model is assumed to be adiabatic. Power for adiabatic compression is calculated by:

$$W_c = m \cdot c_p \cdot \frac{T_i}{\eta_c} \left(\left(\frac{P_O}{P_I} \right)^{\frac{\gamma-1}{\gamma}} - 1 \right) \tag{5.55}$$

where m is the mass flow rate (kg/s); c_p is the specific heat ($J\,kg^{-1}\,K^{-1}$); η_c is the compressor adiabatic efficiency; T_i is the inlet temperature (K); $\frac{P_O}{P_I}$ is the compression ratio and γ is the ratio of specific heat of the gas, $\frac{c_p}{c_v}$.

For air at standard atmospheric pressure and temperature [32], $c_p = 1004\,J\,kg^{-1}\,K^{-1}$ and $\gamma = 1.4$.

Since the adiabatic efficiency is a known value, the power required by the compressor can be evaluated. Of course, like all mechanical devices, there will be mechanical losses while transmitting power from the turbine to the compressor.

5.4.8 Recuperator

The effectiveness of the recuperator is defined as in [31] (Equation 5.56), where $_C$ and $_H$ represent the cold and hot side of the recuperator respectively. The heat exchanger effectiveness is the ratio of the actual heat transfer rate to the maximum possible heat transfer rate if there was an infinite surface area. The heat exchanger effectiveness depends upon whether the hot fluid or cold fluid is a 'minimum fluid', that is, the fluid that has the smaller capacity coefficient as described by:

$$\varepsilon = \frac{T_{c.out} - T_{c.in}}{T_{h \cdot in} - T_{c.in}} \tag{5.56}$$

At design point this is calculated by assuming minimum ΔT is 40 K so that the maximum possible energy is recovered. The recuperator effectiveness at the design point can easily be obtained after the turbine exhaust temperature is calculated. Different approaches based on more realistic behaviour (such as mean differential temperature) may also be used [31].

Table 5.5 Hybrid system assumptions.

SOFC		Combustor pressure loss	5%
SOFC temp (K)	1273	Combustor efficiency	100%
Air utilization	0.3	Recuperator pressure drop	2%
Fuel utilization	0.8		
S/C	3	**Others**	
Current density (A cm^{-2})	0.4	Atmospheric pressure (kPa)	101.325
Pressure loss	3%	Atmospheric temp (K)	288
		Generator efficiency	0.95
Compressor		LHV of the fuel (J/kg)	4.81×10^7
Compressor efficiency	0.75		
Compression ratio	4	**Constants**	
		Faraday's constant (C/mol)	96439
Turbine		R (J mol^{-1} K^{-1})	8.314
Turbine efficiency	0.85	γ of air	1.4
		c_p of air (J kg^{-1} K^{-1})	1004

Source: Adapted from Nien (2007) [16] and Tanaka *et al.* (2000) [29]

The main assumptions used in the hybrid model are presented in Table 5.5.

5.5 Results and Discussion

The hybrid system performance at the design point is shown in Table 5.6. The hybrid system provides an efficiency of 61%, which is higher than conventional means of power generation systems that typically have efficiencies between 30 and 45%. This is because the conventional systems are generally restricted by the fundamental Carnot efficiency, while the SOFC performance is not because the electrochemical energy conversion is a direct power generation process (chemical to electrical).

Moreover, depending on the requirements of the end users, the waste heat might be used in another process, for example in the generation of steam, hot water or hot air. Alternatively, the waste heat can be used for preheating the stream generation unit for the fuel with another recuperator device.

Table 5.6 Estimated system performance of the SOFC-GT hybrid system.

Fuel cell voltage	0.6542 V
Air flow rate	1.8148 kg/s
Fuel flow rate	0.05021 kg/s
LHV of the fuel	4.81×10^7 J/kg
Gas flow LHV	2.415 MW
Total electric power	1.463 MW
System efficiency	60.56%

5.6 Dynamic Models

Performing a dynamic analysis on a hybrid system is another important step to be considered during development. This step is essential to avoid problems (not only at steady-state but also during transient conditions) such as machine overspeed, compressor surge, excessive temperature or thermal gradients in the fuel cell, excessive pressure difference between cathodic and anodic sides, and too low a steam-to-carbon ratio (STCR) on the SOFC anodic side. Dynamic modelling is an essential part in the development of a plant control system.

A number of dynamic models have been developed in the literature, considering a range of different approaches [33–36]. Simplified tools are often necessary to operate time-dependent simulations, especially when analyzing long transient phenomena. 1D/2D models are sometimes included for some components (e.g. the fuel cell), but transient models are typically developed using 0D approaches (just global equations between the component inlet and outlet sections) with 'lumped volume' techniques for the time-dependent responses [33] or considering time constants applied to first-order delay equations, which are considered less reliable because they are not based on real physical input data.

The 'lumped volume' approach is widely used in transient models [33, 37–39], and so details of this approach are presented below. As shown in [37], this technique considers a steady-state off-design model in conjunction with a modelled constant-section pipe (A, L are the equivalent section and length respectively) for the transient response. The equations related to the time-dependent part are: the momentum equation (Equation 5.57) where C is the inlet/outlet pressure difference calculated over time (at steady state it is equal to the value calculated by the off-design model) and the energy equation (Equation 5.58). These equations are coupled with the energy time-dependent calculation (Equation 5.59) for the solid material of the component.

$$\frac{dm}{dt} = \frac{A}{L} \cdot \left[C - \left(p_{out} - p_{t\_in} \right) \right] \tag{5.57}$$

$$\frac{d \left(c_v \cdot \text{mass} \cdot T_t \right)}{dt} = \Delta \left(m \cdot c_p \cdot T_t \right) + q_{solid} \tag{5.58}$$

$$\frac{d(T_{solid})}{dt} = \frac{q_{solid} - q_{lost}}{\text{mass}_{solid} \cdot c_{solid}} \tag{5.59}$$

In case real-time performance is necessary [38], sufficient simplifications to the numerical model must be made to meet this solution speed requirement, whilst bearing in mind that this may have an effect on the model's accuracy and its ability to capture all the relevant physics. Adopting simpler modelling approaches in these cases is often essential. This is especially true for the simulation of long transient phenomena, such as model startup and shutdown procedures, to develop and optimize the control system, to operate hardware-in-the-loop (HIL) analyses, and to perform diagnostic activities on the plant [37].

In comparison with a more complex and time-consuming transient model, the real-time approach has to be based on the following main solutions [37]:

- 0D approaches or component models based on map interpolation;
- A 'lumped volume' technique for the transient part (the momentum equation could be neglected for large timescale calculations);
- A standard interface between components;
- No property calculations in each component;
- No complex subroutines for property calculations (e.g. specific heat has to be calculated with second-order polynomial

functions avoiding equations based on hyperbolic sine and cosine functions.

- Enthalpy balances based on average specific heat values (where possible);
- Minimization of internal iterative calculations;
- Removal of unnecessary components (e.g. data visualization blocks in visual tools).

Since dynamic modelling activity is often linked to control system design and analysis, here we report some details about control systems for SOFC hybrid plants. Due to the complexity of the system, additional controllers have to be included in comparison with a standard turbine plant, because it is necessary to control more properties: SOFC temperature, steam-to-carbon ratio, SOFC current, and so on. Moreover, as already mentioned in Chapter 4, several properties (often not measurable) have to be maintained within narrow limits, such as cathode–anode pressure difference, anodic-side steam-to-carbon ratio, turbine inlet temperature, recuperator temperature, temperature gradients in the fuel cell, and so on. Additionally the compressor must be operated within its surge margin. Considering all these aspects, attention is focused here on SOFC temperature control issues. These present the most significant problem for hybrid system controllers to manage. Several approaches were considered in previous works, and a number of interesting points made on this topic [33, 40–42]:

- Cathodic air (and thus SOFC temperature) may be controlled by regulating the turbine load and corresponding machine rotational speed.
- Controlling the sharing of global power between the turbine and the fuel cell is important to address the necessity of satisfying demand and controlling the turbine speed for SOFC temperature control.
- A compressor–turbine bypass valve can be used to control turbine speed and, as a consequence, SOFC temperature.
- A bleed valve may be used to control the cathode airflow rate by simply discharging a part of compressed air flow (this simple approach is inefficient).

Another problem to be addressed when designing control systems for hybrid systems is that simple PID controllers, typically used in standard systems, are not adequate. This is due to the fact that PID controllers are not able to prevent temperature oscillations during transient operations. As mentioned in Chapter 4, the slow thermal response of the stack and constraints related to temperature gradients must be carefully considered. For this reason, it is usually necessary to use the following more complex approaches: feed-forward [43], model predictive control [44], h-infinity or other innovative control solutions [45].

5.7 Model Validation

Since all the models include assumptions, experimental coefficients or simplified approaches, it is essential to perform a robust validation to assess their performance. The best approach is to perform a detailed comparison between model results and experimental data. However, when the real plant is not available (as is the case during design or development) or the required data are covered by intellectual property issues (a common occurrence with hybrid system modelling activities), a remaining option is to validate individual components or subsystems of the entire plant [46, 47]. A further approach used when experimental/performance data are not available for individual components is the comparison of results with those generated by other validated models. Usually the comparison is made with more detailed models (e.g. using CFD tools) to assess the performance of simplified approaches [48].

A typical validation technique for complex cases, such as for time-dependent models including the control system, can be broken up into the following sub-studies:

- A design point validation requiring all the input values of components (sometimes equivalent values in 0D models) to obtain a good match with experimental or theoretical reference data;
- An off-design validation at steady-state conditions, calculating errors between calculated and reference values to evaluate the model performance and ensure the errors are within an acceptable range;
- A time-dependent validation comparing results with reference data, typically considering simple cases only (e.g. open loop steps not affected by control system performance);

- A validation of the entire model including the control system considering property comparisons in real plant operative conditions.

Following this procedure and meeting the relevant qualifications at each stage results in a well validated model. However, it is important to ensure that the validation is performed over a wide enough operative range, and not only at a few conditions. A recent example of a detailed validation study related to this final aspect is reported in [38] considering a microturbine case for hybrid system applications.

5.8 Conclusion

The appropriate modelling approach and characteristics are very important aspects when developing SOFC-mGT hybrid system models. Due to the plant complexity and the number of interdependent components, it is essential to evaluate a suitable modelling strategy. A significant factor to consider is the performance of the model in terms of its computational work requirement (e.g. while steady-state models can include multidimensional approaches for some components such as the fuel cell, very simple 0D approaches are necessary for transient models, especially if real-time performance is required). To explore the options available, this chapter presented an extended overview of modelling approaches starting from the component level (e.g. GT off-design techniques). An important application of these models is hybrid system gas-turbine matching, an overview of which was presented.

An example of a simple hybrid system steady-state model was presented considering a pressurized layout. This model was carefully studied and results related to the hybrid system performance reported. An interesting efficiency improvement (61% in 1 MW plant size) in comparison with conventional power generation systems was found. The hybrid system model presented in this chapter demonstrates typical approaches used for off-design calculations. The model allows the influence of the different parameters on the plant performance to be investigated.

Finally, attention was focused on dynamic and transient models, including aspects related to real-time solutions and control systems. Due to the importance of this topic, a section related to model validation was included. A range of validation techniques and a robust validation procedure were outlined, given that there is often a need to study and assess a mathematical model performance on plants where experimental data are not yet available (e.g. commercial hybrid system prototypes, or systems in the design or development phase).

5.9 Questions and Exercises

1 What are the common steps for hybrid system model development, and what are the possible ways to reduce complexities/time-dependent problems in simulation activities?

2 Considering the complexities of SOFC-microturbine hybrid systems, explain in detail the approach that involves an engineering procedure based on a mass-matching algorithm, requiring the mass flow leaving the hybrid system to be compatible with that at entry to the turbine.

3 Describe the fundamental techniques for calculating non-equilibrium off-design performance of gas turbines, from simple to complex layouts.

4 How could the performance of hybrid systems be predicted?

5 What are the main modelling solutions to develop a real-time model for hybrid systems?

References

1 Zabihian, F. and Fung, A. (2009) A review on modeling of hybrid solid oxide fuel cell systems. *International Journal of Engineering*, 3 (2), 85–119.

2 Winkler, W., Nehter, P., Williams, M.C., Tucker, D. and Gemmen, R. (2006) General fuel cell hybrid synergies and hybrid system testing status. *Journal of Power Sources*, 159 (1), 656–666.

3 Bove, R., Lunghi, P. and Sammes, N.M. (2005) SOFC mathematic model for systems simulations. Part one: from a micro-detailed to macro-black-box model. *International Journal of Hydrogen Energy*, 30 (2), 181–187.

4 Magistri, L., Bozzo, R., Costamagna, P. and Massardo, A.F. (2002) Simplified versus detailed SOFC reactor models and influence on the simulation of the design point performance of hybrid systems. ASME Turbo Expo 2002: Power for Land, Sea, and Air, pp. 895–903.

5 Judkoff, R.D. and Neymark, J.S. (1995) A procedure for testing the ability of whole building energy simulation programs to thermally model the building fabric. *Journal of Solar Energy Engineering*, 117, 7.

6 Yakabe, H., Ogiwara, T., Hishinuma, M. and Yasuda, I. (2001) 3-D model calculation for planar SOFC. *Journal of Power Sources*, 102 (1), 144–154.

7 Petruzzi, L., Cocchi, S. and Fineschi, F. (2003) A global thermo-electrochemical model for SOFC systems design and engineering. *Journal of Power Sources*, 118 (1), 96–107.

8 Padulles, J., Ault, G.W. and McDonald, J.R. (2000) An integrated SOFC plant dynamic model for power systems simulation. *Journal of Power Sources*, 86 (1), 495–500.

9 Achenbach, E. (1994) Three-dimensional and time-dependent simulation of a planar solid oxide fuel cell stack. *Journal of Power Sources*, 49 (1), 333–348.

10 Peksen, M., Al-Masri, A., Blum, L. and Stolten, D. (2013) 3D transient thermomechanical behaviour of a full scale SOFC short stack. *International Journal of Hydrogen Energy*, 38, 4099–4107.

11 Nikooyeh, K., Jeje, A.A. and Hill, J.M. (2007) 3D modeling of anode-supported planar SOFC with internal reforming of methane. *Journal of Power Sources*, 171, 601–609.

12 Tang, S., Amiri, A., Vijay, P. and Tadé, M.O. (2016) Development and validation of a computationally efficient pseudo 3D model for planar SOFC integrated with a heating furnace. *Chemical Engineering Journal*, 290, 252–262.

13 Ghigliazza, F., Traverso, A., Massardo, A.F., Wingate, J. and Ferrari, M.L. (2009) Generic real-time modeling of solid oxide fuel cell hybrid systems. *Journal of Fuel Cell Science and Technology*, 6, 021312_1-7.

14 Saravanamuttoo, H.I.H., Rogers, G.F.C. and Cohen, H. (2001) *Gas Turbine Theory*. Pearson Education, Harlow, UK.

15 Trasino, F., Bozzolo, M., Magistri, L. and Massardo, A.F. (2011) Modeling and performance analysis of the Rolls-Royce Fuel Cell Systems Limited: 1 MW plant. *Journal of Engineering for Gas Turbines and Power*, 133, 021701-1-11.

16 Nien, C.W. (2007) Design and development of combined fuel cell–gas turbine cycles for environmentally friendly, efficient power generation. MSc thesis, University of Manchester.

17 Uechi, H., Kimijima, S. and Kasagi, N. (2004) Cycle analysis of gas turbine–fuel cell cycle hybrid micro generation system. *Journal of Engineering for Gas Turbines and Power*, 126 (4), 755–762.

18 Costamagna, P., Magistri, L. and Massardo, A.F. (2001) Design and part-load performance of a hybrid system based on a solid oxide fuel cell reactor and a micro gas turbine. *Journal of Power Sources*, 96 (2), 352–368.

19 Costamagna, P., Arato, E., Antonucci, P.L. and Antonucci, V. (1996) Partial oxidation of CH_4 in solid oxide fuel cells: simulation model of the electrochemical reactor and experimental validation. *Chemical Engineering Science*, 51 (11), 3013–3018.

20 Costamagna, P. (1997) The benefit of solid oxide fuel cells with integrated air pre-heater. *Journal of Power Sources*, 69 (1), 1–9.

21 Bessette, N.F., Wepfer, W.J. and Winnick, J. (1995) A mathematical model of a solid oxide fuel cell. *Journal of the Electrochemical Society*, 142 (11), 3792–3800.

22 Chan, S.H., Ho, H.K. and Tian, Y. (2002) Modelling of simple hybrid solid oxide fuel cell and gas turbine power plant. *Journal of Power Sources*, 109 (1), 111–120.

23 Massardo, A.F. and Lubelli, F. (1998) Internal reforming solid oxide fuel cell-gas turbine combined cycles (IRSOFC-GT): Part A – Cell model and cycle thermodynamic analysis. ASME International Gas Turbine and Aeroengine Congress and Exhibition 1998, pp. V003T08A028-V003T08A028.

24 Bossel, U. (1992) *Final report on SOFC Data, Facts and Figures*. Swiss Federal Office of Energy, Operating Agent Task II.

25 Chan, S.H., Low, C.F. and Ding, O.L. (2002) Energy and exergy analysis of simple solid-oxide fuel-cell power systems. *Journal of Power Sources*, 103 (2), 188–200.

26 Gordon, S. and McBride, B.J. (1971) *Computer program for calculation of complex equilibrium composition, rocket performance, incident and reflected shocks and Chapman-Jouguet detonations*. NASA SP-273 168. NASA, US.

27 Burcat, A. (1984) Table of coefficient sets for NASA polynomials, Appendix C. In *Combustion Chemistry*. Springer-Verlag, New York.

28 Burcat, A. (1984) Thermochemical data for combustion calculations. In *Combustion Chemistry*, pp. 455–473. Springer-Verlag, New York.

29 Tanaka, K., Wen, C. and Yamada, K. (2000) Design and evaluation of combined cycle system with solid oxide fuel cell and gas turbine. *Fuel*, 79 (12), 1493–1507.

30 Fukunaga, H., Ihara, M., Sakaki, K. and Yamada, K. (1996) The relationship between overpotential and the three phase boundary length. *Solid State Ionics*, 86, 1179–1185.

31 Damo, U.M. (2016) SOFC hybrid systems equipped with re-compression technology: Transient analysis based on an emulator test rig. PhD thesis, University of Manchester.

32 Energy FC. Inc. (1999) *High Efficiency Fossil Power Plants (HEFPP) Conceptualization Program, Final Report*. Vol. 1 – Technical Report. US Department of Energy.

33 Ferrari, M.L. (2011) Solid oxide fuel cell hybrid system: Control strategy for stand-alone configurations. *Journal of Power Sources*, 196, 2682–2690.

34 Henke, M., Willich, C., Westner, C., Leucht, F., Kallo, J., Bessler, W.G. and Friedrich, K.A. (2013) A validated multi-scale model of a SOFC stack at elevated pressure. *Fuel Cells*, 13, 773–780.

35 Stiller, C., Thoruda, B., Bolland, O., Kandepu, R. and Lars Imsland, L. (2006) Control strategy for a solid oxide fuel cell and gas turbine hybrid system. *Journal of Power Sources*, 158, 303–315.

36 Jia, Z., Sun, J., Oh, S.-R., Dobbs, H. and King, J. (2013) Control of the dual mode operation of generator/motor in SOFC/GT-based APU for extended dynamic capabilities. *Journal of Power Sources*, 235, 172–180.

37 Ferrari, M.L., Traverso, A., Magistri, L. and Massardo, A.F. (2005) Influence of the anodic recirculation transient behaviour on the SOFC hybrid system performance. *Journal of Power Sources*, 149, 22–32.

38 Damo, U.M., Ferrari, M.L., Turan, A. and Massardo, A.F. (2015) Test rig for hybrid system emulation: New real-time transient model validated in a wide operative range. *Fuel Cells*, 1, 7–14.

39 Ghigliazza, F., Traverso, A., Massardo, A.F., Wingate, J. and Ferrari, M.L. (2009) Generic real-time modeling of solid oxide fuel cell hybrid systems. *Journal of Fuel Cell Science and Technology*, 6, 021312_1-7.

40 Henke, M., Willich, C., Westner, C., Leucht, F., Kallo, J., Bessler, W.G. and Friedrich, K.A. (2013) A validated multi-scale model of a SOFC stack at elevated pressure. *Fuel Cells*, 13, 773–780.

41 Stiller, C., Thoruda, B., Bolland, O., Kandepu, R. and Lars Imsland, L. (2006) Control strategy for a solid oxide fuel cell and gas turbine hybrid system. *Journal of Power Sources*, 158, 303–315.

42 Jia, Z., Sun, J., Oh, S.-R., Dobbs, H. and King, J. (2013) Control of the dual mode operation of generator/motor in SOFC/GT-based APU for extended dynamic capabilities. *Journal of Power Sources*, 235, 172–180.

43 Ferrari, M.L. (2015) Advanced control approach for hybrid systems based on solid oxide fuel cells. *Applied Energy*, 145, 364–373.

44 Larosa, L., Traverso, A., Ferrari, M.L. and Zaccaria, V. (2015) Pressurized SOFC hybrid systems: Control system study and experimental verification. *Journal of Engineering for Gas Turbines and Power*, 137, 031602_1-8.

45 Tsai, A., Tucker, D. and Emami, T. (2014) Adaptive control of a nonlinear fuel cell-gas turbine balance of plant simulation facility. *Journal of Fuel Cell Science and Technology*, 11, 061002_1-8.

46 Ferrari, M.L., Traverso, A., Pascenti, M. and Massardo, A.F. (2007) Early start-up of SOFC hybrid systems with ejector cathodic recirculation: Experimental results and model verification. *Proceedings of the Institution of Mechanical Engineers, Part A, Journal of Power and Energy*, 221, 627–635.

47 Ferrari, M.L., Liese, E., Tucker, D., Lawson, L., Traverso, A. and Massardo, A.F. (2007) Transient modeling of the NETL hybrid fuel cell/gas turbine facility and experimental validation. *Journal of Engineering for Gas Turbines and Power*, 129, 1012–1019.

48 Magistri, L., Bozzo, R., Costamagna, P. and Massardo, A.F. (2002) Simplified versus detailed SOFC reactor models and influence on the simulation of the design point performance of hybrid system. ASME Paper 2002-61-30653.

6

Experimental Emulation Facilities

6.1 Experimental Emulation Facilities

Since fuel cell stacks are both very expensive [1] and very susceptible to both thermal and mechanical stresses [2], experimental activities on hybrid systems are regularly carried out using emulator test rigs. These experimental plants are able to produce significant results [3, 4], providing data on specific plant aspects, without the risk to/need for the most critical components, thus reducing the development cost. Emulator rigs enable testing of specific plant components and operational modes (according to the design specifications), but are not able to produce results covering all operational system aspects. For instance, if the fuel cell is not included in the rig, it is not possible to study the electrochemical attributes [5] of the reactor. Emulator rigs are especially useful during the component and control system design stages of the development process, where an appropriate emulator rig is ideally suited for producing the necessary experimental data. There are also a number of other plant operational aspects that can be effectively explored/analyzed with these emulation rigs. Specifically, emulator rigs are extremely useful during

Hybrid Systems Based on Solid Oxide Fuel Cells: Modelling and Design, First Edition.
Mario L. Ferrari, Usman M. Damo, Ali Turan, and David Sánchez.
© 2017 John Wiley & Sons Ltd. Published 2017 by John Wiley & Sons Ltd.

control system design, as during the development of a new control system there is a considerable risk of expensive damage associated with unsuitable operation (e.g. surge [6, 7] or high stress on components [8]). Carrying out any experimental tests with an emulator rig mitigates this factor.

Another important application of emulator test rigs concerns validation/verification activities in support of theoretical models [9, 10]. Specifically, experimental results produced by these facilities can be used to carry out extensive comparative analyses between predicted performance data (obtained from theoretical tools) and the measured values. Furthermore, when it is not possible to complete the validation activity over the entire operational design space (e.g. because certain components are not included in the rig), similitude approaches [11] can be carried out, or the models can be considered for validation via extrapolation techniques [11].

The effective application of a range of different emulator test rigs is employed as a routine developmental tool in many associated research undertakings. For instance, in the renewable energy field a similar emulation approach was used for wind turbines [12], while in the aeronautical field a hardware-in-the-loop (HIL) based test rig was used for studying time-delay effects in power systems [13]. In the mechanical field, emulation techniques were used for developing an experimental platform studying underwater sensor networks [14].

6.2 Reduced-scale Test Facilities

Similarity considerations (e.g. the Reynolds number) [11] can be utilized to allow the use of emulator rigs of a reduced scale. This approach is usually employed when the experimental tests carried out on a reduced-scale rig can be profitably employed for the prototype/actual plant design (e.g. in case of a plant based on a stack of the same components) due to the reduction in facility costs. In order to classify an experimental facility as a reduced-scale emulator, it is necessary to substitute one or more components with a low-cost or low-risk device, but nevertheless retain the usefulness for the required tests. For instance, a reduced-scale fuel cell rig (with a real stack) is not an emulator plant, even if it can be used to carry out significant tests for larger stacks.

In this chapter, attention is focused on specific examples related to reduced-scale emulator rigs. Specifically, the two emulator rigs developed by the TPG (Thermochemical Power Group) of the University of Genoa are presented. This research group has investigated both anodic and cathodic recirculation loops, considering, in both cases, a simple vessel to emulate the fuel cell stack.

6.2.1 Anodic Recirculation Test Rig

This test rig was developed by the TPG in 2003 based on a hybrid system designed by Rolls-Royce Fuel Cell Systems (stack size: 250 kW; fuel utilization: 75%; stack temperatures: 800–970°C; current density: 2940 A/m^2) [15, 16]. For confidentiality reasons, it is not possible to show further details related to the scale reduction used to design the experimental plant. The facility is dedicated to carrying out an experimental analysis of the performance related to ejectors installed in the anodic recirculation of SOFC-based hybrid systems. Specifically, the plant was used to improve the anodic ejector design considering not only the inherent aspects, but also evaluating the impact of the performance modifications related to the various interactions with the anodic circuit (pressurization, pressure losses, high-temperature operation and chemical composition effects).

Figure 6.1 shows the layout [17, 18] for this anodic recirculation rig. It is based on a single-stage ejector fed by compressed gas (usually air, but tests with different gases are also possible) on the primary line. A 100 litre vessel is used to emulate the anodic volume. The ejector outlet is connected to the volume inlet using a manual valve (necessary for pressure loss change) and a volume outlet pipe is connected to the ejector secondary inlet duct. As in a real SOFC anodic side, the vessel also includes a second outlet line equipped with a manual valve. This second valve is necessary to pressurize the system up to 5 bar. An electrical heater was introduced inside the vessel to reach secondary inlet temperature values up to 573 K with thermal insulation. Since the rig was not designed to operate at the real ejector operational conditions normally present in hybrid systems [11], similitude conditions were considered. The results for other similitude

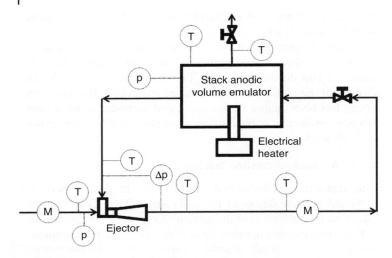

Figure 6.1 Layout of the anodic recirculation test rig.

parameters, including the Reynolds number, were discussed in other earlier studies [19]. As shown in Figure 6.1, the test rig was equipped with different probes to measure the mass flow rate, pressures and temperatures in the main sections of the circuit [11, 17, 18]. It is essential to measure ejector properties carefully, and Figure 6.1 shows the required probes installed in both the primary and secondary ducts to achieve this. While the primary line mass flow rate can be measured in a small pipe (10 mm size) without specific problems, the ejector outlet flow is measured with a Venturi-based probe installed in a 50 mm (nominal diameter) pipe. For this reason, it was necessary to install the device downstream of a straight pipe in order to have a uniform flow, and to equip the probe with a temperature acquisition device. The pressure value is obtained from other probes, as the pressure losses in this pipe can be considered to have a negligible effect on the mass flow rate measurement.

This rig was used in a number of other studies [11, 17, 18] to measure the performance of different anodic ejectors at steady-state conditions. As shown in [11], special attention was devoted to the ejector characteristic curves (pressure increase versus recirculation ratio) measured at both ambient and high-temperature conditions at different vessel operational

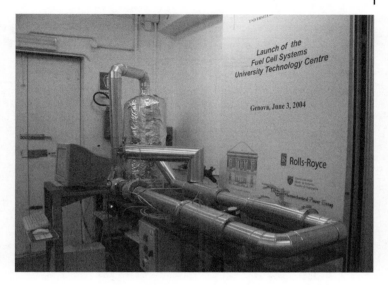

Figure 6.2 Picture of the anodic recirculation test rig.

pressure levels. These performance tests were carried out with air in both the primary and secondary ducts, with CO_2 in the primary line and air in the secondary one, and using CO_2 in all the plant components. This fluid change was necessary, as anodic ejectors in SOFC anodic circuits are significantly affected by chemical composition [11]. The emulator test rig (see Figure 6.2) was essential for performing transient tests involving the ejector and the whole anodic system. These tests were used to validate predictive tools [17, 18] developed to provide realistic simulations regarding complete hybrid system performance [8, 20].

6.2.2 Cathodic Loop Test Rig

To analyze the initial part of a hybrid system startup process, the University of Genoa designed and installed an emulator test rig to retain the highest flexibility, allowing for several different startup configurations. A schematic of the physical simulator of a fuel cell hybrid system can be seen in Figure 6.3. The cathode side utilizes recirculation [21] carried out by a single-stage ejector.

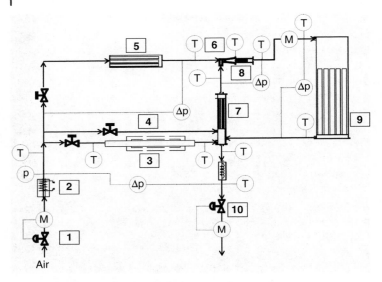

Figure 6.3 Layout of the cathodic loop test rig.

As with the anodic recirculation test rig, this test rig was based on the hybrid system designed by Rolls-Royce Fuel Cell Systems [15, 16]. It was scaled down such that it used a fraction of the total mass flow rate used by the prototype hybrid system. The main goal of this facility is the physical simulation of the conditions related to the initial part of the startup phase (initial 15 minutes), when the stack components are chemically inactive, but already integrated with the airflow path. During this operation, the turbine is still far from self-sustaining conditions, and the flows have to be carefully managed to avoid thermal shocks or backflows.

Since during these operations the fuel cell is chemically inactive, the stack thermal capacitance is emulated with a plenum filled with ceramics (no electrochemical reaction materials). Moreover, the turbine is not actually installed in the rig, but its effects are simulated by a compressor–turbine model [21–23] based on performance maps and rotational mechanical inertia. The turbine emulation is specifically based on the following approach: the rig probes provide the compressor outlet pressure and the expander inlet pressure and temperature to the machine model; the model then calculates and provides the compressor outlet flow and temperature, and turbine exhaust mass flow rate to the rig [21–23]. The compressor behaviour is obtained with a mass flow meter-control device (point 1) and with an electric heater (point 2). This approach generates a very high level of flexibility because it is possible to change the 'virtual' turbine by using different characteristic curves in the real-time model.

Three air paths, managed by automatic on/off valves, are present in the rig. The first one is the startup combustor line (point 3), which is emulated by a high-temperature electrical heater: it is placed on a bypass line or upstream of the expander. A bypass circuit (point 4), which is the second path, is included. The third path is the cathodic ejector primary nozzle line (point 6). The facility is equipped with physical emulators of thermal and volume capacitance [22]. The vessel located upstream of the cathodic primary nozzle (point 5) is necessary to emulate the capacitance between the compressor outlet duct and the ejector primary inlet line; the volume installed at the startup combustor outlet section (point 7) is a mixing device equipped with its capacitance; the ejector plenum (point 8) is necessary to emulate the capacitance between the ejector outlet duct and stack inlet pipe; the fuel cell vessel (point 9) represents the stack volume and capacitance. The system discharge is managed with a servo-controlled valve, which is necessary to emulate the expander (point 10). No anodic side is currently included in the facility because the cell is chemically inactive during this early startup phase.

The startup combustor is based on a stainless-steel pipe [24] surrounded by two ceramic semi-shells for two 3.5 kW electrical heaters (point 3) [22, 24]. This device is thermally insulated with ceramic material and rock wool [24]. The layout of this device allows it to reach around 1000°C (1273.15 K) air outlet temperature, as necessary to emulate a real startup combustor. Realistic combustor dynamic behaviour is correctly obtained with the bypass branch. Specifically, the electrical heater is warmed up to the operative temperature before beginning the test and the bypass line is used to generate the real time-dependent behaviour of the combustor outlet temperature trend [22].

The facility is equipped with probes and a data acquisition system: LabVIEW^{TM} software, which interacts via LAN with a

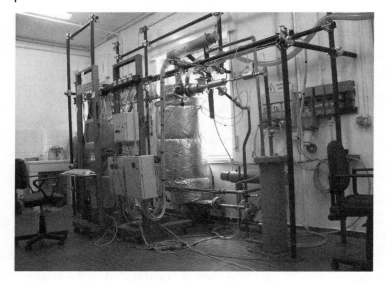

Figure 6.4 Picture of the cathodic loop test rig.

FieldPoint^{TM}. Figure 6.4 shows a picture of the facility including all the transducers used to operate the tests.

Tests were carried out to emulate the initial 15 minutes of the startup phase using generic characteristic maps of the turbomachinery [22, 23]. These tests were carried out at ambient temperature or using the electrical heaters. In the second case, the electrical heater installed for emulating the startup combustor was managed such that the air outlet temperature of this device was close to 950°C (1223.15 K) [22]. The results obtained during these tests were used to verify the mass flow rate split between each line, the ejector recirculation performance and the component thermal response. A wide validation of a time-dependent model developed with the TRANSEO tool [25] at the University of Genoa was carried out in different operative conditions [22].

6.3 Actual-scale Test Facilities

These emulator rigs are based on actual-scale components. This approach is used to improve the emulation performance reaching operational conditions for realistic plants. However, to reduce costs and avoid risk conditions, some components are emulated with simple devices. A typical example of this approach is shown by the fuel cell emulation. Since this component is usually expensive and extremely critical from a thermal, mechanical and chemical stress point of view, a typical emulation approach consists of substituting this component with a same-sized vessel and a combustor (in the case of high-temperature rigs).

Our attention focuses on examples related to actual-scale emulator rigs; both low- and high-temperature facilities are presented.

6.3.1 Low-temperature Rigs

These facilities are simple emulation rigs designed to carry out specific tests. Since they are not able to reach the specific temperature conditions of an actual fuel cell, they cannot provide a complete hybrid system emulation. However, they are usually designed and installed to perform tests on components for which the performance of each can be considered significant at low-temperature conditions. For instance, tests on compressor surge risks can be carried out at cold conditions (without an actual combustor) if the interaction with the other components is produced with a vessel and a valve located downstream of the compressor.

The following low-temperature emulation facilities are presented: the surge test rig [26] developed by the University of Trieste, Italy, and the emulation rig for tests on control components [27] designed and installed by the ESTIA Institute of Technology, France.

6.3.1.1 Surge Test Rig

This test rig is composed of a compressor, a 3.9 m³ pressure vessel, an expander and a valve [26]. This facility is only able to perform 'cold tests' because no combustors or fuel cells are installed. For this reason, the turbine is powered by a variable-speed electrical motor coupled to the turbine through a gearbox (20:1 gear ratio). The facility layout is shown in Figure 6.5. The gas turbine is an experimental prototype able to operate at 60,000 rpm, 0.7 kg/s and producing a 3.5 compression

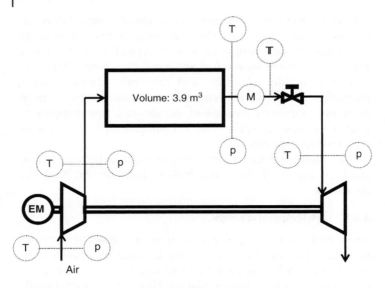

Figure 6.5 Layout of the surge test rig.

ratio. The compressor-vessel pipe (90 mm inner diameter) is composed of a 1.6 m steel line and a 9.7 m flexible hose, while the vessel-expander pipe has a 7 m steel line followed by a 4.6 m flexible hose. This layout was chosen to optimize the available spaces and without any consideration of possible real plant layout and size. The mass flow rate is measured with a diaphragm installed at the volume outlet. A butterfly valve was installed in this rig (see Figure 6.5) to operate tests with different pressure loss values between the compressor outlet and turbine inlet. The data acquisition system is carried out at 500 Hz sampling time and the measured values are stored with a PC-based system.

The experimental tests were carried out considering 'deceleration profiles' [26]. Starting from steady-state conditions the electrical motor was switched off. These tests carried out at different butterfly valve positions showed several pulsations related to surge conditions. Other conditions leading to surge were analyzed to highlight possible risky operations to be avoided in SOFC-based hybrid systems.

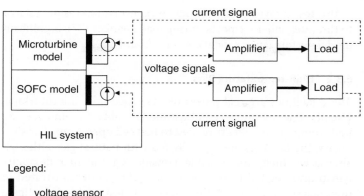

Legend:

▮   voltage sensor

⊕   controlled current source

----▸   signal

━▸   amplified signal

Figure 6.6 Layout of the control test rig.

6.3.1.2   Emulation Rig for Tests on Control Components

This test rig is based on a few hardware components, including the interface between load resistor banks and a software-based simulator, connected using a hardware-in-the-loop device to run a hybrid system real-time model. As shown in Figure 6.6 (developed on the basis of [27]), all the plant components are simulated with a model developed in a Matlab®-Simulink® environment working on the OPAL-RT real-time platform. Since this model has to operate in a real-time environment, simple approaches are used. For instance, the dynamic behaviour of specific components is represented using transfer functions, avoiding the application of complex time-dependent equations. The output signals of this model are amplified using two 1 kW four-quadrant amplifiers. The microturbine power output value is converted to a DC signal. When the hybrid system model operates at the design power (30 kW for the microturbine and 90 kW for the SOFC), the power absorbed by load devices is 100 times lower. The amplifiers are controlled by the models of the two energy sources.

This test rig was used to produce experimental results related to time-dependent aspects during load changes [27]. Significant model performance validation was obtained [27].

6.3.2 High-temperature Rigs

These facilities are real power plants based on a microturbine. Usually one or more combustors are able to emulate the high-temperature effects related to fuel cell operations. For simplicity, the facilities are typically fed with natural gas, although alternative fuels are possible considering the fuel flexibility constraints related to microturbines [28]. As demonstrated in previous studies [29, 30], these test rigs, designed to investigate hybrid system aspects (performance, component integration, transient operations, risk conditions, control system issues, etc.) without expensive SOFC stacks, are an interesting solution for producing experimental results [29] relevant to the running of entire realistic plants.

In this chapter the following high-temperature emulation facilities are presented (in chronological order): the emulator rig developed by the US Department of Energy's NETL (West Virginia) [29, 31], the emulation facility developed by the University of Genoa (Italy) [32], and the test rig for hybrid system emulation by the DLR (Germany) [30].

6.3.2.1 Emulator by the US Department of Energy – NETL

The emulator rig developed by NETL is the first worldwide facility to perform emulation tests on SOFC hybrid systems with an actual-scale plant (including a real gas turbine). It was developed and installed in the early 2000s to operate tests on the coupling between an SOFC stack and a microturbine, focusing special attention on surge risk prevention and control system aspects [31, 33–35]. The emulator rig consists of a fuel cell emulator, composed of a natural gas burner and a couple of vessels, coupled to a recuperated microturbine. The application of this device in the initial phase of the system design process allows the analysis of transient events and control system issues without the risk of destroying a real fuel cell stack in case of unfavourable operations (e.g. surge events). The microturbine (see Figure 6 7) is a 120 kW auxiliary power unit for aircraft, composed of a single shaft turbine, a radial compressor (design values: pressure ratio of 4 and 2 kg/s air mass flow rate) and a gear-driven electrical generator (load produced with a resistor bank).

Downstream of the two counterflow heat exchangers (the recuperators), a 2 m^3 stainless-steel vessel is installed to emulate the stack cathode volume behaviour. This vessel is connected to the combustor (necessary to generate the same thermal effect as the SOFC and to sustain the machine) located upstream of a 0.6 m^3 vessel (designed to emulate the off-gas burner size) and the expander. The turbine outlet duct is connected to the recuperator hot side. The facility also includes a bleed line and two bypass ducts managed by control valves. These devices are essential for carrying out tests for specific conditions encountered by hybrid systems (e.g. the bleed line is essential to avoid surge during the shutdown and other phases), for control system development and for good emulation of transient phases.

The rig is equipped with probes to measure significant design and development parameters. Rotational speed is obtained with an optical sensor, and pressures and temperatures are measured with appropriate probes located upstream and downstream of the plant main components. The rig also includes differential pressure sensors (not shown in Figure 6.7 for the sake of clarity) to measure the pressure losses for the most significant components [33]. The air mass flow rates are measured with annubar flow meters at the compressor inlet, at the fuel cell vessel inlet, and in the cold bypass line. Another mass flow meter is used for the natural gas. The control system of the rig was completely redeveloped by Woodward Governor and the NETL staff [34]. Further details related to this test rig are reported in several previous works [31, 33–35].

The emulation of a complete hybrid system is obtained with the coupling of the two vessels with the combustor chamber fed by natural gas (Figure 6.7). Since this is an emulator plant and the actual SOFC stack is not present in this rig, the real thermal behaviour is obtained with a real-time model used to drive the fuel valve. This real-time model includes the fuel cell, the reformer, the off-gas burner and the anodic recirculation system [34]. A software-in-the-loop approach (or hardware-based simulations) is used to generate the model results (produced

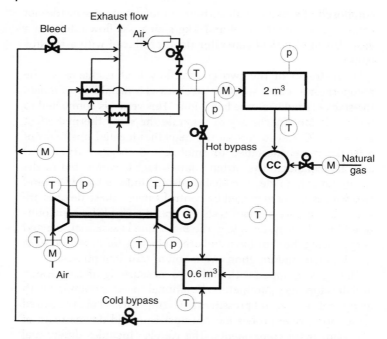

Figure 6.7 Layout of the emulator by the US Department of Energy – NETL.

in real-time mode), which are used to manage the fuel valve to match the measured turbine inlet temperature with the calculated one.

This test rig was used to study the SOFC/mGT matching aspects [36], focusing special attention on surge prevention [37], ambient pressure effects [33] and control system development [38]. The facility was used to study different control strategies from the standard PID approach [38] to innovative solutions [39, 40] using in-house control devices.

6.3.2.2 Emulator by the University of Genoa – TPG

The hybrid system emulator rig developed by the University of Genoa – Thermochemical Power Group (TPG) was installed during the 2006–2011 period. The general layout is similar to that of NETL, in that it is a recuperated microturbine connected to one or more vessels to emulate the stack dimension. The main difference between the two is that the test rig in Genoa is a hybrid system emulator, while at NETL the plant couples a turbine and a fuel cell emulator. An important difference between the rigs is related to the simulated part: while at NETL the software is able to directly control the plant fuel flow rate, in the test rig by TPG the hardware/software coupling is obtained by generating the matching between calculated and measured turbine outlet temperature values. Moreover, the test rig by TPG also includes the following aspects [32, 41–43]:

- the rig is based on a commercial machine connected to the fuel cell emulation devices with simple modifications;
- standard components are used (e.g. the control system and machine case are not modified);
- the facility includes a real ejector-based anodic side (the vessel and recirculation system);
- the cathodic vessel is modular to carry out tests considering SOFC stacks of different sizes or dimensions;
- the rig is equipped with a water system able to manage the control of the compressor inlet temperature (using water from the supply system or cooled with an absorption chiller);
- the facility is equipped with a steam injection system to emulate chemical composition aspects of fuel cell outlet ducts;
- a recompression system is included for specific tests.

This hybrid system emulator test rig is based on the coupling of a T100 commercial recuperated microturbine with a device for fuel cell emulation. This system [41–43] comprises a cathodic side modular vessel [41], located between the recuperator outlet and combustor inlet, and an anodic circuit [43] (equipped with a second vessel). The test rig layout is shown in Figure 6.8 and the related picture in Figure 6.9. The machine is a Turbec T100 PHS Series 3 [44]. This machine is equipped to operate in a stand-alone configuration or connected to the electrical grid. It works at a constant rotational speed if the turbine is in stand-alone mode. In grid-connected mode the machine controller works at constant turbine outlet temperature (TOT). This turbine is composed of a complete module [44] for power generation (nominal values: $100\,kWe$, $N = 70000\,rpm$, $\beta = 4.45$), a heat exchanger (WHEx) located downstream of the recuperator outlet (hot side) for cogenerative applications. When the machine is in stand-alone mode the startup phase

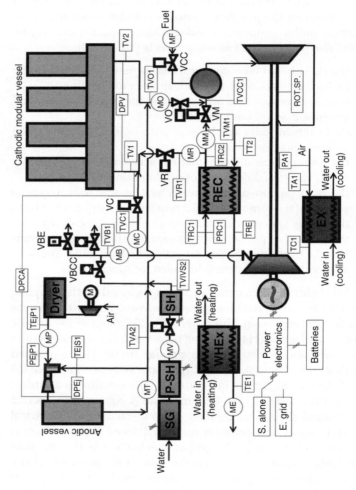

Figure 6.8 Layout of the emulator by the University of Genoa – TPG.

Figure 6.9 Picture of the emulator by the University of Genoa – TPG.

is performed using two battery packages. The power unit is modified in order to couple with the vessels by way of a small modification to the pipe layout, as shown in [45]. The machine is equipped with a compressor inlet temperature control system [46], an essential device to manage this temperature value. This facility is based on three air/water heat exchangers installed at the machine air intakes (Ex in Figure 6.8) and connected to the water system. Several pipelines are included for different purposes [32]: the vessel bypass duct managed by the VM valve, the recuperator bypass line managed by the VC valve, the recuperator/vessel bypass line managed by the VBCC line, a controlled bleed line managed by the VB valve, and the lines to connect the cathodic vessel to the machine (managed by VR and VO valves). Water flow from the supply line or from an absorption chiller (see [47] for more details) is able to produce the required temperature values (a constant value, or a trend typical of ambient temperature variation).

The cathodic modular vessel emulates the fuel cell cathodic volume, as shown in [32]. It is composed of two collector pipes, connected at the recuperator outlet and the combustor inlet respectively, and four module pipes connected to both collectors. Both collectors and module pipes have a nominal diameter of 350 mm and their total length is about 34 m. The maximum volume is about 3.2 m^3. This modular structure allows the emulation of components suitable for the machine size, such as fuel cells with a planar or tubular structure. The maximum volume is related to an SOFC size of about 380 kW. It is based on a similitude approach [48] with the RRFCS stack (see [15, 48–50] for further SOFC details).

The anodic recirculation system, designed on the basis of the RRFCS hybrid system (scaled up to 450 kW) [49], is composed of a compressed air line (to emulate the fuel flow with an air mass flow rate up to 20 g/s [49]), an anodic single-stage ejector, and a suitable vessel for the anodic volume emulation. This is a useful approach to emulate the fuel mass flow rate at the ejector primary duct with an airflow. This line is composed of a 15 kW compressor for air, a dryer, and the following sensors: mass flow rate (MP in Figure 6.8), pressure (PEjP1), and temperature (TEjP1). The ejector generates the recirculation flow rate as in a typical SOFC hybrid system. The anodic volume is a 350 mm diameter pipe (U pipe) for a total volume of about 0.8 m^3. It was designed using the aforementioned similitude approach with the RRFCS fuel cell (see [15, 48–50] for details on this SOFC).

Since a hybrid system exhaust flow has a higher percentage of steam and CO_2 than a standard gas turbine, to emulate the expander inlet flow, two different approaches are possible: c_p or k $= c_p/c_v$ similitude conditions. The physical emulation strategy considered for this emulator facility is the c_p similitude condition. To simplify the injection process, only the steam injection system is included, and the CO_2 effect is obtained with an additional mass flow rate of steam. Implementing the steam injection system introduces a number of modifications to the base system, starting with the installation of a steam generator (SG) able to produce at least 27 g/s mass flow rate, a superheater (SH) to increase the steam temperature to a temperature suitable for the turbine combustor inlet (about 515°C), and a controlled (VIV) valve for the mass flow rate management. Since the measuring of the steam flow rate requires the steam to be completely mono-phase, an additional electrical heater (P-SH) is installed.

Since in systems based on a cathodic ejector a significant pressure drop is present in the fresh air line, it is necessary to operate the plant with a two-stage machine [50], or using both a standard single-shaft microturbine and a commercial recompression device [51]. To operate emulation tests with recompression devices, a tee joint is included to drive the airflow from the T100 compressor outlet to an intercooler. This device is essential to decrease the air temperature (with water flow from an evaporative tower) to reduce the power absorbed by the second compression. The intercooler outlet duct (air side) is connected to the compressor inlet of a turbocharger (a Garrett GT3067). To generate the pressure loss related to the cathodic ejector primary duct [51], an electronically controlled valve is installed in the facility. The power necessary to operate the recompression is obtained via the high-temperature flow driven from the vessel outlet to the GT3067 turbine inlet. More details on these additional modifications to the facility (including the plant layout related to this machine coupling) are given in [51].

To manage the facility with all the additional instrumentation and control valves shown in Figure 6.8, a new acquisition and control system has been developed using the LabVIEW™ software. It runs coupled with the machine standard software [44] to ensure safe machine operations.

This facility was (and is) essential to perform different tests on hybrid system technology. Initially, several results were produced on SOFC/turbine coupling issues, starting from surge prevention (e.g. using the bleed valve during shutdown operations) and moving on to an extensive analysis of the system dynamics, including the emulation of the startup/shutdown phases of the entire hybrid system. Several results produced in these phases confirmed the conclusions previously obtained by US DOE/NETL (e.g. methods of surge prevention during system transient operations). Then, special attention was focused on individual plant components at both steady-state and dynamic conditions (e.g. an extensive experimental campaign on the recuperator or on the anodic ejector). The additional components, beyond those present in the NETL plant, allow for additional important experimental activities to be carried out. These additional capabilities allowed for studies addressing the following topics using the specified additional components in both steady-state and transient modes: interaction between anode and cathode side [43], influence of SOFC outlet chemical composition on turbine performance [42], and pressure increase with a recompressor device [51, 52]. Special attention was also devoted to control system aspects using a real-time model for components not physically present in the facility. As with the work carried out by NETL, the model was used to generate the effect of fuel cell outlet flow on the machine. However, since in this plant it is not possible to control the fuel flow rate directly with a valve (the fuel valve is managed by the T100 control system), the hardware/software coupling was obtained by matching the measured turbine outlet temperature with the related calculated value (changing the electrical load in stand-alone mode). Several interesting results were obtained using both traditional [53] and innovative [54] control logic systems.

6.3.2.3 Emulator by the DLR

The DLR (Deutschen zentrum für Luft- und Raumfahrt) started the development of their hybrid system emulator rig in 2008. The general layout of this facility is similar to that of NETL and TPG (a recuperated microturbine connected to a vessel for stack dimension emulation). More specifically, the following aspects give an overview of the facility layout with respect to the facilities by NETL and TPG [30]:

- Machine aspects similar to the TPG layout (a commercial machine connected to the fuel cell emulation devices with simple modifications);
- Control system aspects similar to the TPG layout (standard components);
- An anodic side similar to the NETL layout (not included);
- A cathodic vessel similar to the TPG layout (variable volume);
- A compressor inlet temperature management similar to the NETL layout (no related hardware was installed).

In contrast to both previous test rigs, the DLR emulator plant is able to provide chemical composition effects with greater accuracy as the real turbine inlet mass/molar fractions can be obtained with an additional combustor located inside the vessel (Figure 6.10). Another important aspect, in comparison to the

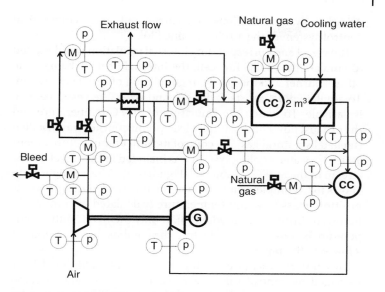

Figure 6.10 Layout of the emulator by the DLR.

plants by NETL and TPG, relates to the vessel temperature. Specifically, in the emulator rig developed by DLR the entire vessel can reach actual fuel cell temperature values, as the temperature control is carried out with a cooling water flow.

The facility comprises a commercial microturbine (Turbec T100 PH), additional pipes for subsystem matching, and an SOFC emulator (a $2\,m^3$ vessel). The test rig was designed to have the highest flexibility: the gas turbine can operate in a standard recuperated layout or with the additional piping and the vessel. With this approach the facility allows testing of the influence of the additional components on the microturbine. Since the gas turbine [44] is very similar to that presented for the TPG test rig, the reader is directed to the previous section for specific details. Some components of the machine, however, were modified as discussed below. The lean premix combustor was configured to have the possibility of including optical devices. The pipes between the compressor outlet and recuperator inlet were modified to include a coriolis mass flow probe. To incorporate these modifications the machine case was completely removed. A new cooling circuit was installed for oil and water cooling.

Additionally, modifications were carried out on the commercial control system of the T100 machine [30].

The vessel and corresponding internal insulation was designed to emulate the SOFC fuel cell, the internal heat exchanger, the off-gas burner and the anodic recirculation of a fuel cell system. To emulate different stack designs and sizes, the vessel volume is variable. To match the emulation of SOFC temperature and the exhaust flow composition, a lean premix burner was installed (fuelled by natural gas). To control its outlet temperature (about 900°C), a water-cooling system was included in the vessel downstream of the SOFC combustor (Figure 6.10). This configuration allows for a complete emulation of the real stack outlet gas composition, mass flow and temperature (calculated by a model in hardware-in-the-loop mode) [30]. If only the SOFC outlet temperature is considered for the test, the combustor can operate at a lower outlet temperature.

This test rig was also equipped with three additional lines managed by control valves: a bleed line for preventing surge operations, a recuperator bypass duct for startup tests, and a vessel bypass line to operate the T100 turbine in its standard configuration. The rig was equipped with several instrumentation devices (thermocouples, pressure transducers, coriolis mass flow meters and a humidity sensor). Specifically, 56 thermocouples were connected to a Delphin data acquisition system, and 12 total and 12 static pressure transducers were coupled to a pressure scanner module (from Esterline). The ambient conditions (static pressure, temperature and humidity) are measured close to the compressor inlet duct. The exhaust gas composition can be determined by an analyzer manufactured by ABB. This device can detect the concentration of O_2, CO, CO_2, NO, NO_2 and unburned hydrocarbons. All signals related to these probes are sent via Ethernet to the data-logging software developed in the LabVIEW^{TM} environment [30].

This test rig was designed to analyze the dynamic interaction between the different subsystems, focusing attention on the effects on the gas turbine. This ability makes the rig a vital tool in the design, test and validation of new control systems for the gas turbine and for the entire hybrid system [30]. At the moment, the facility is used for model validation [55] and for preliminary tests on the fully instrumented turbine [56]. Special attention is devoted to surge prevention aspects, including measurements on the performance curves of the turbomachines [56].

6.4 Conclusions

This chapter has presented a comprehensive description/discussion regarding the characteristics of widely employed emulator rigs and their use in hybrid system design and development activities. Although, depending on the specific aim of a facility development, a wide range of components and layouts are possible, the chapter classifies plants (hybrid system emulator rigs) into broad categories, and discusses the types of problems analyzed using each of these rigs, with a discussion and overview of the relevant related engineering data. Starting from reduced-scale facilities and moving to full-scale power plants equipped with an actual microturbine, the chapter presents the primary results obtained with these plants, which are of great importance to activities ranging from component integration issues to solutions for control system aspects. Related references are included for the reader to find additional details for each plant discussed. A significant improvement in the pace of research undertakings related to such technology issues must be highlighted, even though the commercial significance of applying the conclusions reached by studies using these emulation plants will only have a major effect when hybrid systems eventually reach mature commercialization levels. Furthermore, it is important to state that these experimental research activities based on emulation facilities allow for a significant reduction in costs by a reduction in the risk of component damage during tests with complete prototypes.

6.5 Questions and Exercises

1   How can the emulator test rigs for SOFC hybrid systems be classified?

2 Which emulation approach can be used to study a plant component (e.g. the anodic ejector)?

3 Which emulation approaches can be used to study the plant control system?

4 Which emulation approaches can be used to study the stack thermal response?

5 In an emulator test rig composed of a T100 microturbine coupled with a cathodic vessel (main layout similar DLR's plant) it is necessary to evaluate the maximum pressure loss that can be included between the recuperator outlet (cold side) and the combustor inlet for a minimum electrical power of 20 kW. Evaluate this pressure loss considering the following data:

- Compressor inlet temperature: 15°C (ambient temperature)
- Compressor inlet pressure: 1.013 bar (ambient pressure)
- Air filter pressure loss: 5 mbar
- Compressor adiabatic efficiency: 0.79
- Compressor pressure ratio: 4.5
- Recuperator (cold side) pressure loss (including the connection ducts): 80 mbar
- Combustor pressure loss: 2.5% of the inlet pressure
- Turbine adiabatic efficiency: 0.86
- Turbine outlet temperature: 645°C
- Recuperator (hot side) pressure loss (including ducts): 7% of the ambient pressure
- Air mass flow rate: 0.78 kg/s
- Generator and power electronic efficiency: 0.85
- Power absorbed by auxiliaries: 10 kW.

6 In a SOFC emulator vessel (cylindrical layout) it is necessary to evaluate the thermal insulation (the thickness for a rock wool layer) considering that the external layer is an aluminium cover (1 mm thickness) and that the external temperature must be retained at 50°C at steady-state conditions. Consider the following data (neglecting the contact thermal resistances between the different materials):

- Ambient temperature: 15°C
- Ambient pressure: 1.013 bar
- Vessel inner temperature: 600°C
- Vessel inner pressure: 4 bar
- Vessel inner diameter: 1 m
- Vessel thickness: 5 mm
- Vessel conductivity: 105 W/(mK)
- Rock wool layer conductivity: 0.08 W/(mK)
- Aluminium layer conductivity: 117 W/(mK).

References

1 Franzoni, A., Magistri, L., Traverso, A. and Massardo, A.F. (2008) Thermoeconomic analysis of pressurized hybrid SOFC systems with CO_2 separation. *Energy*, 33, 311–320.

2 Soudarev, A.V., Konakov, V.G., Molchanov, A.S., Souryaninov, A.A. and Tikhoplav, V.Y. (2005) Increasing efficiency and reliability of hybrid engines (SOFC+μGTE) operation by hydrogen separation from a high temperature co-containing flow using molecular ceramic membranes. *Proceedings, Electrochemical Society*, PV 2005-07, 240–248.

3 Shelton, M., Celik, I., Liese, E. and Tucker, D. (2010) A study in the process modeling of the startup of fuel cell/gas turbine hybrid systems. *Journal of Engineering for Gas Turbines and Power*, 132, 012301_1–8.

4 Tsai, A., Tucker, D. and Emami, T. (2014) Adaptive control of a nonlinear fuel cell-gas turbine balance of plant simulation facility. *Journal of Fuel Cell Science and Technology*, 11, 061002_1–8.

5 Chen, G., Kishimoto, H., Yamaji, K., Kuramoto, K. and Horita, T. (2015) Effect of interaction between A-site deficient LST and ScSZ on electrochemical performance of SOFC. *Journal of the Electrochemical Society*, 162, F223–F228.

6 Fink, D.A., Cumpsty, N.A. and Greitzer, E.M. (1992) Surge dynamics in a free-spool centrifugal compressor system. *Journal of Turbomachinery*, 114, 321–332.

7 Arnulfi, G.L., Giannattasio, P., Giusto, C., Massardo, A.F., Micheli, D. and Pinamonti, P. (1999) Multistage centrifugal

compressor surge analysis: Part II – Numerical simulation and dynamic control parameters evaluation. *Journal of Turbomachinery*, 121, 312–320.

8 Ferrari, M.L. (2015) Advanced control approach for hybrid systems based on solid oxide fuel cells. *Applied Energy*, 145, 364–373.

9 Damo, U.M., Ferrari, M.L., Turan, A. and Massardo, A.F. (2015) Test rig for hybrid system emulation: New real-time transient model validated in a wide operative range. *Fuel Cells*, 1, 7–14.

10 Shelton, M., Celik, I., Liese, E., Tucker, D. and Lawson, L. (2005) A transient model of a hybrid fuel cell/gas turbine test facility using Simulink. ASME Turbo Expo 2005, Reno-Tahoe, NV, USA.

11 Ferrari, M.L., Pascenti, M. and Massardo, A.F. (2008) Ejector model for high temperature fuel cell hybrid systems: Experimental validation at steady-state and dynamic conditions. *Journal of Fuel Cell Science and Technology*, 5, 041005_1–7.

12 Tammaruckwattana, S., Ohyama, K. and Yue, C. (2015) Experimental assessment with wind turbine emulator of variable-speed wind power generation system using boost chopper circuit of permanent magnet synchronous generator. *Journal of Power Electronics*, 15, 246–255.

13 Gan, C., Todd, R. and Apsley, J. (2013) *Time-Delay Effects in a HIL Aircraft Power System Emulator*. IEEE Energy Conversion Congress and Exposition, Denver, CO, pp. 20–26.

14 Ying, W., Zhigang, J. and Yishan, S. (2014) Simulator-to-emulator: Analysis and design of experiment platform for underwater sensor networks. *Applied Mechanics and Materials*, 473, 219–225.

15 Agnew, G.D., Bozzolo, M., Moritz, R.R. and Berenyi, S. (2005) The design and integration of the Rolls-Royce Fuel Cell Systems 1MW SOFC. ASME Turbo Expo 2005, Reno-Tahoe, NV, USA.

16 Trasino, F., Bozzolo, M., Magistri, L. and Massardo, A.F. (2011) Modeling and Performance Analysis of the Rolls-Royce Fuel Cell Systems Limited: 1 MW Plant. *Journal of Engineering for Gas Turbines and Power*, 133, 021701_1–11.

17 Ferrari, M.L., Bernardi, D. and Massardo, A.F. (2006) Design and testing of ejectors for high temperature fuel cell hybrid systems. *Journal of Fuel Cell Science and Technology*, 3, 284–291.

18 Ferrari, M.L., Pascenti, M. and Massardo, A.F. (2006) Experimental validation of an unsteady ejector model for hybrid systems. GT2006-90447, ASME Turbo Expo 2006, Barcelona, Spain.

19 Costabile, L. (2005) *Experimental Analysis of anodic and cathodic recirculations for solid oxide fuel cell based hybrid systems*. MS thesis, University of Genoa, Italy (in Italian).

20 Ferrari, M.L. (2011) Solid oxide fuel cell hybrid system: Control strategy for stand-alone configurations. *Journal of Power Sources*, 196 (5), 2682–2690.

21 Traverso, A., Ferrari, M.L., Pascenti, M. and Massardo, A.F. (2006) Physical simulator of start-up for hybrid system cathode side. FUELCELL2006-97051, ASME Fuel Cell Conference, Irvine, California, USA.

22 Ferrari, M.L., Traverso, A., Pascenti, M. and Massardo, A.F. (2007) Early start-up of SOFC hybrid systems with ejector cathodic recirculation: Experimental results and model verification. *Proceedings of the Institution of Mechanical Engineers, Part A, Journal of Power and Energy*, 221, 627–635.

23 Traverso, A., Ferrari, M.L., Massardo, A.F. and Pascenti, M. (2009) Turbovir: hardware simulation of virtual turbomachinery. Italian Patent, deposit no. 0001359369.

24 Traverso, A., Ferrari, M.L., Massardo, A.F. and Pascenti, M. (2009) Non intrusive heater for fluids. Italian Patent, deposit no. 0001361914.

25 Traverso, A. (2005) TRANSEO code for the dynamic simulation of micro gas turbine cycles. ASME Paper 2005-GT-68101.

26 Taccani, R. and Micheli, D. (2006) Experimental test facility for the analysis of transient behavior of high temperature fuel cell/gas turbine hybrid power plants. *Journal of Fuel Cell Science and Technology*, 3, 234–241.

27 Baudoin, S., Vechiu, I., Camblong, H., Vinassa, J.-M., Barelli, L. and Kreckelbergh, S. (2014) Analysis and validation of a biogas hybrid SOFC/GT emulator. IEEE International Workshop on Intelligent Energy Systems (IWIES). doi:10.1109/IWIES.2014.6957053

28 Taamallah, S., Vogiatzaki, K., Alzahrani, F.M., Mokheimer, E.M.A., Habib, M.A. and Ghoniem, A.F. (2015) Fuel flexibility, stability and emissions in premixed hydrogen-rich gas turbine combustion: Technology, fundamentals, and numerical simulations. *Applied Energy*, 154, 1020–1047.

29 Shelton, M., Celik, I., Liese, E. and Tucker, D. (2010) A study in the process modeling of the startup of fuel cell/gas turbine hybrid systems. *Journal of Engineering for Gas Turbines and Power*, 132, 012301_1–8.

30 Hohloch, M., Widenhorn, A., Lebküchner, D., Panne, T. and Aigner, M. (2008) Micro gas turbine test rig for hybrid power plant application. ASME Paper GT2008-50443.

31 Tucker, D., Liese, E., VanOsdol, J., Lawson, L. and Gemmen, R.S. (2003) Fuel cell gas turbine hybrid simulation facility design. ASME International Mechanical Engineering Congress and Exposition, Proceedings, pp. 183–190.

32 Ferrari, M.L., Pascenti, M., Bertone, R. and Magistri, L. (2009) Hybrid simulation facility based on commercial 100 kWe micro gas turbine. *Journal of Fuel Cell Science and Technology*, 6, 031008_1–8.

33 Tucker, D., Lawson, L., VanOsdol, J., Kislear, J. and Akinbobuyi, A. (2006) Examination of ambient pressure effects on hybrid solid oxide fuel cell turbine system operation using hardware simulation. ASME Paper No. 2006-GT-91291.

34 Shelton, M., Celik, I., Liese, E., Tucker, D. and Lawson, L. (2005) A transient model of a hybrid fuel cell/gas turbine test facility using Simulink. ASME Paper No. 2005-GT-68467.

35 Ferrari, M.L., Liese, E., Tucker, D., Lawson, L., Traverso, A. and Massardo, A.F. (2007) Transient modeling of the NETL hybrid fuel cell/gas turbine facility and experimental validation. *Journal of Engineering for Gas Turbines and Power*, 129, 1012–1019.

36 Tucker, D., Lawson, L. and Gemmen, R. (2003) Preliminary results of a cold flow test in a fuel cell gas turbine hybrid simulation facility. ASME Paper GT2003-62074.

37 Pezzini, P., Tucker, D. and Traverso, A. (2013) Avoiding compressor surge during emergency shutdown hybrid turbine systems. *Journal of Engineering for Gas Turbines and Power*, 135, 102602_1–10.

38 Tucker, D., Lawson, L. and Gemmen, R. (2005) Characterization of air flow management and control in a fuel cell turbine hybrid power system using hardware simulation. Proceedings of the ASME Power Conference, PWR2005-50127, pp. 959–967.

39 Tsai, A., Tucker, D. and Clippinger, D. (2011) Simultaneous turbine speed regulation and fuel cell airflow tracking of a SOFC/GT hybrid plant with the use of airflow bypass valves. *Journal of Fuel Cell Science and Technology*, 8, 061018_1–10.

40 Tsai, A., Tucker, D. and Emami, T. (2014) Adaptive control of a nonlinear fuel cell-gas turbine balance of plant simulation facility. *Journal of Fuel Cell Science and Technology*, 11, 061002_1–8.

41 Ferrari, M.L., Pascenti, M., Magistri, L. and Massardo, A.F. (2010) Hybrid system test rig: Start-up and shutdown physical emulation. *Journal of Fuel Cell Science and Technology*, 7, 021005_1–7.

42 Ferrari, M.L., Pascenti, M., Traverso, A.N. and Massardo, A.F. (2012) Hybrid system test rig: Chemical composition emulation with steam injection. *Applied Energy*, 97, 809–815.

43 Ferrari, M.L. and Massardo, A.F. (2013) Cathode-anode interaction in SOFC hybrid systems. *Applied Energy*, 105, 369–379.

44 Turbec T100 Series 3 (2002) *Installation Handbook*. Turbec, Malmö, Sweden.

45 Pascenti, M., Ferrari, M.L., Magistri, L. and Massardo, A.F. (2007) Micro gas turbine based test rig for hybrid system emulation. ASME Paper GT2007-27075, ASME Turbo Expo, Barcelona, Spain.

46 Ferrari, M.L., Pascenti, M., Magistri, L. and Massardo, A.F. (2010) Micro gas turbine recuperator: Steady-state and transient experimental investigation. *Journal of Engineering for Gas Turbines and Power*, 132, 022301_1–8.

47 Ferrari, M.L. and Pascenti, M. (2011) Flexible micro gas turbine rig for tests on advanced energy systems, in *Advances in*

Gas Turbine Technology (ed. INTECH). Rijeka, Croatia, pp. 89–114.

48 Ferrari, M.L., Pascenti, M., Magistri, L. and Massardo, A.F. (2011) MGT/HTFC hybrid system emulator test rig: Experimental investigation on the anodic recirculation system. *Journal of Fuel Cell Science and Technology*, 8, 021012_1–9.

49 Ferrari, M.L., Pascenti, M., Magistri, L. and Massardo, A.F. (2009) Hybrid system emulator enhancement: Anodic circuit design. ASME Paper ICEPAG2009-1041, International Colloquium on Environmentally Preferred Advanced Power Generation 2009, Newport Beach, California, USA.

50 Trasino, F., Bozzolo, M., Magistri, L. and Massardo, A.F. (2011) Modeling and performance analysis of the Rolls-Royce Fuel Cell Systems Limited: 1 MW plant. *Journal of Engineering for Gas Turbines and Power*, 133, 021701_1–11.

51 Ferrari, M.L., Pascenti, M., Traverso, A. and Massardo, A.F. (2015) Re-compression system for SOFC hybrid plants: Tests with an emulator rig. International Gas Turbine Congress, Tokyo.

52 Damo, U.M., Ferrari, M.L., Turan, A. and Massardo, A.F. (2015) Re-compression model for SOFC hybrid systems: Start-up and shutdown test for an emulator rig. *Fuel Cells*, 1, 42–48.

53 Caratozzolo, F., Ferrari, M.L., Traverso, A. and Massardo, A.F. (2013) Emulator rig for SOFC hybrid systems: Temperature and power control with a real-time software. *Fuel Cells*, 6, 1123–1130.

54 Larosa, L., Traverso, A., Ferrari, M.L. and Zaccaria, V. (2015) Pressurized SOFC hybrid systems: Control system study and experimental verification. *Journal of Engineering for Gas Turbines and Power*, 137, 031602_1–8.

55 Henke, M., Klempp, N., Hohloch, M., Monz, T. and Aigner, M. (2015) Validation of a T100 micro gas turbine steady-state simulation tool. ASME Paper GT2015-24090, ASME Turbo Expo 2015.

56 Zanger, J., Widenhorn, A. and Aigner, M. (2010) Experimental investigations of pressure losses on the performance of a micro gas turbine system. ASME Paper GT2010-22067, ASME Turbo Expo 2010.

7

Problems and Solutions for Future Hybrid Systems

In this chapter, attention will be devoted to potential hybrid system problems and possible solutions analyzed using theoretical/computational and experimental tools, including physical systems and related available information from commercial companies. Special attention will be given to the role of these kinds of power systems in a future generation paradigm based on distributed generation concepts, using renewable sources and hydrogen as an energy carrier. Micro gas turbine and fuel cell (mGT/SOFC) coupling will be presented, showing potential solutions to size and pressure issues and the utilization of existing technology with more innovative devices. Moreover, the solutions developed for recirculation (both at anode and cathode side) will be presented and discussed. Special attention will be focused on dynamic and control system problems, showing different potential solutions to prevent failures arising from the large volume of the stack or from a

Hybrid Systems Based on Solid Oxide Fuel Cells: Modelling and Design, First Edition.
Mario L. Ferrari, Usman M. Damo, Ali Turan, and David Sánchez.
© 2017 John Wiley & Sons Ltd. Published 2017 by John Wiley & Sons Ltd.

variety of considerations involving vital design constraints. Also, innovative control techniques based on multiple property management will be considered to reduce thermal stress or other property oscillations that can reduce component life.

Due to the importance of alternative fuels for hybrid systems in terms of cost decrease and/or emission issues, attention is focused on future perspectives for hydrogen generation from coal gasification and on biofuels applied to hybrid systems. Due to the importance of environmental issues, CO_2 sequestration is also discussed.

7.1 The Future of Micro Power Generation Systems

The future evolution of hybrid systems involving SOFC/mGT is to a large extent linked to the development of small-size (of the order of 1 MW and less) power generating units in the distributed generation paradigm [1]. Currently, the adoption of such distributed generation systems is increasing due to undeniable benefits [2] (mainly decrease in transportation losses, enablement of cogeneration and flexibility increase) in comparison with the centralized generation paradigm. Most of these systems share common features (low emission performance is essential), and even if different technologies can be used, the most mature choices for the prime movers are, at the moment, microturbines [3] and reciprocating engines [4]. These standard technologies are foreseen to play an important role due to low costs (reciprocating engines), low maintenance (microturbines), low emissions (if fuel is natural gas) and high global efficiency (with cogeneration or trigeneration). Renewable sources [5] will also be easily integrated into distributed generation systems due to their simplicity, easy installation and proximity to the users (e.g. solar panels on building roofs) and very low or zero-emission performance. For these reasons, high-temperature fuel cells [6] and hybrid systems [7, 8] are envisaged as strong candidates for future distributed power systems. These technologies simultaneously hold a number of desirable features: highest electrical efficiency, small/scalable units based on fossil fuels or biofuels, and low gaseous/acoustic emissions. Additionally, hybrid systems are not only exclusively designed for high-efficiency fossil fuel utilization, but they enable the utilization of renewable energy sources like biofuels. Hybrid systems are a perfect complement for a wider system featuring hydrogen generation, storage and utilization at high efficiency [9].

Micro generating power systems pose very interesting research, development and construction challenges, many of which are related to complex thermodynamic aspects [10], chemistry of materials, manufacturing and optimization issues. In this latter regard, distributed power generation involves the development/operation of smart polygeneration grids based on multiple power generation sets (different types and technologies) and consumers, whose management can be related to very different performance metrics (in terms of cost, efficiency and emissions). In essence, the optimizer must take into account the availability of renewable energy sources, the cost of fossil fuels and the load demand distribution in order to satisfy this with the optimum share amongst the different generators (operating with different efficiencies, emissions, and cost aspects) [11]. These optimization tools will require extensive experimental tests, both during development and for validation purposes, before being implemented in a distributed grid, in order to ensure that they are reliable optimizers able to control these systems in real-time mode. To this end, tests based on experimental facilities will be necessary both for software validation and to develop control logics [12, 13] (at both the prime mover or grid levels) and optimization algorithms [14, 15].

As a candidate to be incorporated into these smart grids, SOFC hybrid systems have to be carefully included in these optimization analyses considering all their specific features, such as the potential cost decrease (expected to be significant due to improvements in materials, layout and manufacturing), the different performance trends at part-load conditions (in comparison with traditional fossil-fuel plants based on heat engines), and any other aspect that arises from experimental activities [15, 16] (starting from the initial prototypes) and devoted to integration in polygeneration smart grids.

7.2 The Future of Hybrid Systems: Hydrogen as an Energy Carrier

Most developed/industrialized countries are intensively investing in fuel cells and hybrid system technologies [17]. These activities are mainly aimed at using hydrogen fuel both in transportation, and in large stationary power generation units, which normally use fossil-derived fuels [17]. However, many technical issues must be resolved before fuel cells can attain a reasonable market penetration. The specific technologies must be made cheaper, by conducting robust research and development to compete with the internal and external combustion engines that have, in some cases, already attained good efficiencies and low costs [17].

Hydrogen produced from coal gasification is the US Department of Energy's primary source in the fuel programme, with a primary objective being to develop modules for producing hydrogen from coal at prices competitive with crude oil, when integrated with advanced coal power systems [18]. This program will be important for the development of hydrogen as an energy carrier/vector and, at the moment, involves research activities from both academic and industrial sectors [16, 19]. Researchers involved in these topics are also working on the integration of renewable sources into a hydrogen-based future economy. This approach would be essential to compensate for the 'high oscillations' related to renewable sources through hydrogen generation (and storage) during periods where the available renewable energy sources are not utilized fully [16]. In spite of the advantages in producing near-zero harmful emissions and the interest shown in such development, the general consensus of the scientific community is that the widespread use of hydrogen as a fuel in the foreseeable future appears to be doubtful due to the high energy demands of its production, and other issues related to safety, storage, and distribution [17]. Hence, dedicated wide-ranging research activities are planned over the upcoming decades involving both theoretical and experimental investigations. The development of a hydrogen-based economy is potentially realizable only based on technological and economic improvements related to several subsystems: fuel processing reactors, fuel cells, hybrid systems, hydrogen storage devices, and so on.

7.2.1 Hydro-methane and Hydrogen-rich Fuel Mixtures

Other critical aspects concern final usage of hydrogen, given the high cost and technical issues of fuel cells and other hydrogen-powered systems [20]. The conversion of hydrogen into other fuels or 'chemicals', such as hydro-methane, is considered an interesting solution to bypass these cost issues operating with standard prime movers. For this reason, several research activities, such as the IDRO-RIN TRAN-GENESI project [16, 20], are devoted to this field. Hydro-methane is produced using H_2 and CO_2, according to the well-known Sabatier reaction [20]:

$$4H_2 + CO_2 \rightarrow CH_4 + 2H_2O \qquad (7.1)$$

The reaction is very exothermic and progresses in a temperature range between 250–400°C at fairly low pressures (2–10 bar) on a catalytic bed of nickel or ruthenium on an Al_2O_3 substrate [20]. The gas obtained from the process, after water separation, comprises primarily methane, hydrogen and unreacted carbon dioxide; by definition, hydro-methane (H_2/CH_4) is a gas mixture composed usually of CH_4 and containing from 5% to 30% of H_2 by volume (from 0.6% to 5.4% by mass). Its energy content is close to that of natural gas and much higher than the bio-gas produced from biomass anaerobic digestion [20] (which normally comprises 50–55% CH_4 and 40–45% CO_2), and it can be stored and used applying the same technologies already developed for natural gas. Furthermore, current studies have justified that limited amounts of hydrogen (not greater than 30%) have positive effects on natural gas-based reciprocating engines or piston engines, increasing thermodynamic efficiency and decreasing CO_2, CO and HC emissions, without needing significant modifications in the engines [20]. For these reasons, H_2/CH_4 generation could potentially provide a very attractive solution to storing energy for different applications, in particular for transportation, employing hydro-methane in traditional CH_4-fuelled systems [20].

Researchers at the University of Stuttgart Institute of Thermodynamics and Thermal Engineering have been concentrating on the effect of hydrogen-rich natural gas mixtures on pressurized SOFC systems [21]. Given that the internal reforming steps act as a heat sink, with less natural gas to be reformed, the cells would have to be cooled with different solutions. This would ultimately lead to higher parasitic losses due to the higher air supply. Usually, the control system requires information on the heating value of the fuel gas, so with changing hydrogen content in the fuel, measuring or generating this information for fuel flow control would become crucial [21]. The research group's results show that hydrogen-rich fuels can be used in SOFC systems, even if they have a high influence on the performance and operating temperatures. So, these aspects have to be carefully taken into account by the plant control system [21].

7.3 Future Hybrid Systems: Design, Optimization and Sizing

Although most hybrid system prototypes are based on natural gas, solutions fuelled by pure hydrogen (produced from electricity generated by renewable energy systems) [16] or biogas [22] are also of interest. Specifically, high efficiency considerations related to hybrid systems are responsible for the substantial interest in this technology integrated with renewable sources (as stated previously, to offset renewable source intermittency).

Due to the variability related to renewable sources, energy can be provided by the grid when the plant is not able to meet the load demand. When excess energy is produced during low local demand conditions, such as at night in the case of a wind plant, the surplus energy can be sold to the grid [23], although the feasibility of this option is disputable. For instance, this is feasible if the share of renewables in the grid is low, and thus there is always a chance to export renewable electricity to the grid in spite of very low demand. If the share of renewables is already too high, it will not be possible to export to the grid more renewable electricity than the total actual demand. Another scenario where the viability of this approach is uncertain is in remote areas with an absence of an electrical grid, where the high connection cost (due to large distances and irregular topography) often leads organizations to explore alternative solutions. Stand-alone hybrid systems are considered as one of the most promising enabling technologies to handle the electrification requirements of these regions [24]. Targeting this market, and due to the higher complexity related to the control system development for hybrid systems operating in a stand-alone mode, several research activities are taking place in this field [25].

As for any power generation technology, the selection of hybrid system components is an important aspect during the design and development stage. This ensures reliable and stable electricity supply at an acceptable (and optimized) price [26, 27]. Hence an optimal sizing method is highly recommended for the efficient and economic utilization of renewable energy sources [24]. The optimal sizing of such systems is carried out via analysis for each intended location, which is important because of the influence of site-dependent variables such as energy demand, site temperature and component costs [24, 26]. Computer-based simulation and optimization have received more attention recently, and have become an important consideration in the design of power systems requiring detailed analysis [28, 29].

7.3.1 Hybrid Systems Sizing Techniques

As for any power generation technology, the sizing of hybrid systems is challenging due to factors such as power system optimal design, system type and location, as these must be considered in order to produce a solution that meets the level of demand at reasonable costs [30].

The design of hybrid systems can be determined by the energy demand, their overall fixed [23] and variable costs, levels of emissions and reliability. This information is then combined using standard tools in project appraisal. The net present worth (NPW) or net present value (NPV) can be defined as the difference between the present value of cash inflows and the present value of cash outflows; therefore, the hybrid system with the highest NPV is regarded as the 'optimal design' [31, 32]. NPV is nevertheless an uncertain value that should not be used on its own to make an investment decision. Rather, it must be used in combination with other financial figures of merit such as the internal rate of return (IRR: the discount rate for which $NPV = 0$; the optimal system is that with highest IRR if this turns out to be higher than the required rate of return), the payback period (time to recover the capital invested on the project) and, quite frequently in the context of power generation, the 'levelized cost of electricity' (net present cost of each kWh produced over the lifetime of the power plant). The optimal design can be enhanced by adopting many techniques. Several optimization techniques have been employed by researchers for the sizing of hybrid systems (especially for the technology to be integrated with renewable sources). The following section will provide details of these techniques.

7.3.2 Hybrid System Sizing Simulation Tools

It is common practice for researchers to employ commercially available software (or in-house tools at a commercial level of maturity) to provide a reasonable analysis of the particular hardware design and development problem. For the sizing or selection of optimal designs of hybrid systems, many software packages can be used for free or bought online.

HOMER is one of the best-known simulation packages designed for hybrid renewable micro grids, whether remote or attached to a larger grid. HOMER optimization and sensitivity analysis algorithms allow the designer to evaluate the economic and technical feasibility of a large number of technology options and to account for variations in technology costs and energy resource availability. This software was initially designed by the National Renewable Energy Laboratory in the US for the Village Power Program, and is now owned by HOMER Energy [32]. This sizing tool is capable of carrying out hybrid system simulation, optimization and sensitivity analysis. It can be used for both conventional and renewable energy technologies.

The structure of HOMER is based on the following points:

- *Power sources*: solar photovoltaic cells (PVs), wind turbines, run-of-river hydropower, power generators fed by both traditional and alternative fuels, co-fired electric utility grid, microturbines and fuel cells [33].
- *Storage*: flywheels, battery banks, flow batteries, hydrogen.

- *Loads*: daily profiles with seasonal variation, deferrable (water pumping, refrigeration), thermal (space heating, crop drying), efficiency measures.

Hybrid2 [34] is a computer simulation model for hybrid power systems. Hybrid power systems are designed for the generation and use of electric power. They are independent from large, centralized electricity grids and incorporate more than one type of power source. They may range in size from relatively large island grids to individual household power supplies. This hybrid system computer model, developed at the University of Massachusetts and the National Renewable Energy Laboratory, is a comprehensive, flexible, user-friendly model that allows a wide range of choices of system components and operating strategies.

Optimization of Renewable Intermittent Energies embodied in hydrogeN for auTonomous Electrification (ORIENTE) [35] is a software package aimed at simulating systems using renewable energy sources, and is currently enjoying widespread interest. ORIENTE can simulate and optimize the behaviour of a renewable energy system (RES) coupled with a hydrogen chain (H_2) consisting of an electrolyser (EL), a fuel cell (FC), several storage systems for gas and water, a gas compressor to power loads that meet the constraint of autonomy, along with other forms of storage (batteries, etc.).

W-TEMP (Web ThermoEconomic Modular Program) [36, 37] was designed almost ten years ago when the Thermochemical Power Group (TPG) in Genoa decided to generate a visual version of their in house TEMP program, for the study and optimization of complex and innovative energy systems. W-TEMP allows the thermoeconomic and exergoeconomic analysis of a large number of energy cycles to be obtained involving items such as steam and gas turbines, combined and advanced cycles (mixed gas–steam cycles, biomass gasification integrated plant, fuel cells – SOFC and MCFC – and hybrid cycles, partial oxidation cycles, chemical recovery cycles, integrated solar combined cycles, etc.).

W-ECoMP (Web-based Economic Cogeneration Modular Program) [16, 36] is a modular and flexible software tool, also developed by TPG, which focuses on the optimization of energy systems, including two calculation levels starting from system size definition to optimization of plant management considering load demand values. W-ECoMP is characterized by a modular approach and a standard component interface, which allows the user to build complex cycle configurations in a short time. This approach maintains the flexibility and extendibility of library components. Each component is described by three subroutines, computing mass and energy flows, off-design performance curves, variable and capital costs.

Another piece of simulation software is named Grid-connected Renewable Hybrid Systems Optimization (GRHYSO) [38, 39]. It is worthwhile to refer to [40] for a more comprehensive study on commercially available software tools for the performance evaluation of hybrid renewable energy systems, optimization techniques for hybrid renewable energy system sizing, and possible promising techniques for future use in hybrid system sizing.

7.4 Cost Analysis of Hybrid Systems for Power Generation Applications

In spite of the fact that hybrid systems have attracted a lot of interest in the last decade from the research community and industry, they have not reached commercial maturity yet. This is due to several problems that can be classified into two distinct groups: economic and technology issues. Furthermore, the thermoeconomic analysis of such hybrid SOFC systems is essential, and raises considerations related to individual component manufacturing, maintenance costs and cost reductions for wider market penetration. For instance, considering microturbine design and development aspects, costs of hardware, software, manuals and training, the initial cost of a micro turbine is around €620–980 per kilowatt (a further 30–50% may be added for installation costs). However, turbine manufacturers are trying to reduce such costs and keep them below €620 per kilowatt for the future, considering component cost decreases specifically (including power electronics and control hardware) and leveraging on the benefits derived from economies of scale (i.e. mass production) [41]. Nonetheless, when considering fuel cells, the primary drivers that come to mind are the costs of fuel production, fuel infrastructure, and materials/manufacturing.

These costs can be analyzed from the perspective of cost of materials and manufacturing, fuel, fuel processing and infrastructure and, as a result, the cost of the generated electricity. A huge impediment against fuel cell penetration from conventional energy sources could also be removed with improvements in efficiency (in addition to reliability), especially in case of the application of low-cost fuels (e.g. coal) [41]. Thus, current fuel cell technology is cost-prohibitive in its capability to challenge/compete economically with alternative technologies for most applications. In the numerous niche applications where considerations of efficiency, emissions and noise are paramount (e.g. space or military applications), and/or with substantial subsidies, fuel cell systems are both economically and technologically attractive, and have demonstrated adequate performance metrics [41].

For SOFC-based hybrid systems, material costs have to be carefully taken into account as these often overshadow the whole system cost. It is, however, undeniable that mass production and economies of scale will contribute much towards lowering system costs [41]. In the following discussion, a brief review of the state of the art is presented, including aspects or analyses regarding hybrid system costs.

In [29], thermodynamic and economic analyses were carried out for distributed generation power plants. Specifically, particular attention was devoted to thermoeconomic models of three hybrid system plants with and without fuel decarbonization and carbon dioxide sequestration. The results show the thermodynamic and economic impact of the fuel pre-treatment and CO_2 separation and compression [29].

In [42], a study on the coupling of a SOFC reactor with a small gas turbine puts special emphasis on highlighting the economic considerations/concerns regarding the system (SOFC-mGT). The study entailed a detailed 2D numerical model of a tubular SOFC reactor integrated into the thermoeconomic modular program named TEMP [31]. The authors presented the thermodynamic and electrochemical results obtained for the specific system at the design point. The study presented, in conclusion, a detailed thermoeconomic discussion regarding aspects for the cost of electricity (COE) produced via the SOFC-mGT generator, based on the SOFC cost equation [42].

A further research activity was presented in [43] involving both an exergy and a thermoeconomic analysis of internal reforming solid oxide fuel cell (IRSOFC) and gas turbine (GT) combined cycles using the TEMP tool. A suitable equation for IRSOFC cost evaluation based on cell geometry and performance was proposed and employed to evaluate the cost of generating electricity with the proposed combined systems. The results included a discussion on the influence of several parameters such as external reformer operating conditions, fuel-to-air ratio, cell current density, compressor pressure ratio, and so on [43]. The cost of the electricity in pressurized systems (efficiency values better than 70% were obtained) that was calculated in [43] was significantly lower than that of the atmospheric ones.

Another research group [44] from Montana State University conducted a cost analysis of hybrid systems. This was carried out after the unit sizing had been completed. The data used was for a typical household in the Pacific Northwest, scaled to represent 500 residences [45], and for each configuration the annualized cost and a cost of electricity (COE) were given. The performance characteristics of the microturbine and SOFC power plant used in this evaluation were reported in [44]. This included a Capstone C200 microturbine, rated at 200 kWe, interconnected with 5 kWe SOFC modules; for detailed explanation and information about the 5 kWe SOFC model, the reader is directed to [44, 45].

Regarding the SOFC stack, the individual cells were anticipated to be replaced multiple times during the project lifetime (20 years), while the remainder of the components would last for the entire project [44]. The results show that SOFC stack costs must decrease to $175/kW by 2020 and the delivered cost of H_2 must reach $1/gasoline gallon equivalent (gge) by 2017 in order to reach a competitive threshold [44]. In conclusion, based on current capital costs for the hybrid systems, the analysis demonstrates a distinct economic advantage for the conventional electrical generation methods already in place [45].

Various researchers have reported detailed economic analyses to estimate cost contributions of different hybrid system components and manufacturing processes, with different and sometimes conflicting conclusions. However, all such studies show cost reduction to be absolutely essential for a realistic potential future commercial development of SOFC-based hybrid systems. The research activities that are under development for cost decrease targets are focusing on the following engineering issues:

- Material cost decrease primarily for the fuel cell stack and the reformer;
- Material cost decrease for pressure vessels and pipes related to the application of low- or intermediate-temperature SOFC technology;
- Manufacturing cost decrease using appropriate cell geometry and high-density packaging (e.g. planar layout instead of the tubular one);
- Cost decrease using low-cost manufacturing processes;
- Component cost decrease using reliable and simple solutions (e.g. ejectors for recirculations);
- Plant cost decrease considering innovative layouts (e.g. cathodic recirculation can substitute for the expensive recuperator);
- Cost decrease avoiding expensive sensors or control solutions;
- Cost decrease related to the optimization of power electronics, controlling hardware and auxiliary devices;
- Cost decrease related to series production of all the plant components.

Several authors have analyzed the aforementioned cost decrease aspects regarding experimental development efforts, presenting advantages and disadvantages of each. These issues are addressed in detail by Rolls-Royce Fuel Cell Systems in a study focusing particularly on the development of a hybrid system prototype able to meet a significant cost decrease target for the entire plant [46]. However, since, no hybrid system prototypes are currently able to reach cost reduction targets such that they are widely commercially viable, several further research activities have to be carried out to arrive at a meaningful engineering conclusion.

7.5 Performance Degradation Problems in Solid Oxide Fuel Cells

Performance degradation has a huge effect on high-temperature fuel cells, from electrode delamination and electrolyte cracking, to electrode poisoning and microstructure coarsening; hence such considerations will eventually affect the coupling of SOFCs with micro gas turbine technology [47–50]. Degradation problems cause a decrement in power output at constant current and can even result in the catastrophic failure of the entire cell. The natures and sources of these problems are invariably different, and many models have been developed to study different types of degradation, determine their causes and look for potential solutions [51]. It is important to state that SOFC degradation has an important impact on the possible market penetration of hybrid systems, as achieving longer lifetime (accompanied by high performance) could eventually compensate for higher component costs in comparison with traditional plants. Indeed, despite SOFCs having a high efficiency and being flexible in terms of fuel usage, making them an attractive technology for future energy generation, their particular economic uncompetitiveness is still a major drawback. This is particularly true for the problems related to their short lifetime, due to the multiple degradation phenomena [51].

Concerned about such issues, the National Energy Technology Laboratory (NETL) of the US Department of Energy carried out research activities [51] on key fuel cell parameters during cell degradation using a distributed model. Degradation was taken into account as an increment in ohmic resistance, as a function of the following operating parameters: current density, fuel utilization, and temperature. The choice of these parameters was due to the effects that they have on cell degradation, and because they can be measured and controlled during the operations in order to minimize degradation effects. The use of a simple algebraic expression allows monitoring of the real-time performance of the model [51].

The results showed that the cell (fed by syngas) begins degradation immediately at the inlet, where current density and fuel utilization are higher and the temperature is lower.

This increases the resistance at the inlet and shifts the current density peak toward the exit of the cell. After 500 hours at a constant total current, fuel flow and airflow, the current density and fuel utilization are higher in the centre of the cell. After 1000 hours, temperature decreases at the inlet, with a subsequent negative impact on the local degradation rate, and increases at the outlet, keeping the resistance and the degradation rate low. After 2000 hours, the current density peak is shifted once again to the cell outlet where the resistance is lower and increases significantly as a function of time. After around 12,000 hours, the degradation rate is fairly uniform along the cell [51]. An analysis of localized degradation effects shows how the different parts of the cell degrade at different rates, and how the various profiles are redistributed along the cell as a consequence of different degradation rates [51].

Since the cited analyses show the importance of reducing SOFC degradation to reach the market penetration target, it will be necessary to continue such research activities during the development of hybrid systems. Thus, it is essential to devote the required research/engineering efforts to the following main points (a strong collaboration with industrial teams is essential to have wide access to experimental data):

- Identification of possible operative conditions to reduce degradation rates;
- Identification of possible compromises between performance and degradation issues;
- Material and manufacturing improvements to decrease degradation rate;
- Lifecycle solutions to decrease degradation rate (e.g. decrease in number of startup/shutdown phases);
- Degradation analyses with low-temperature cells to identify possible benefits.

7.6 Turbomachinery Problems

Several authors, including [52], studied the influence of turbomachinery problems on the overall performance of hybrid systems, also considering issues related to material choice for these components. These analyses focused on the external conditions influencing the general requirements for hybrid system performance optimization in a distributed generation application: fuel (natural gas), location (office buildings, hospital, large stores, military bases), and size (0.1–7 MW power output). The layout and design parameters of such a system were determined (at maximum efficiency), leading to the optimization of the system and the assessment of the impact on turbomachinery performance requirements. Off-design operation of the system was analysed showing the appropriate range of turbomachinery performance parameters [52], along with the system configuration and the fuel cell specific parameters for given electrolyte material and thickness. As expected, the parameters and type of turbomachinery required is dependent on the choice of electrolyte [52]. Moreover, as the turbomachinery subsystem only contributes 10–20% of the total power output, an analysis of the control strategy for the integrated system shows that reasonable working conditions can be achieved by maintaining the fuel utilization factor as high as possible and keeping a constant stack temperature; this strategy ensures highest system efficiency [52].

Despite a number of improvements and achievements highlighted in past literature, the matching between the SOFC and the turbine remains an open issue to be improved with further research activities. The main aspects to be considered are as follows:

- As turbomachinery flows are fully unsteady, strictly three-dimensional and turbulent, the off-design conditions are dominated by extremely complicated aero-physics, which is now beginning to be explored via targeted and validated computational fluid dynamic (CFD) predictions and experiments [53]. It is expected that such activities will assume further prominence as design and development using such tools will improve the performance of subsystems and the coupling between them; such considerations are absolutely essential for the optimization of integrated performance, especially off-design in terms of characterization of losses and other design issues [53, 54].

- We require the development of microturbines with working cycle and turbomachinery designs that are tailored to the requirements of hybrid systems. This implies a simultaneous optimization of both systems, gas turbine and fuel cell, for which a large market has to be developed first.

- The development and standardization of gas bearings for the microturbine is needed [55]. Also, the development of both high-speed turboalternators, which at the moment, are expensive due to the limited availability in the market [56], and lower-cost recuperators for microturbines [57].
- The dynamic characteristics of hybrid system operation can, in certain conditions, run the compressor into surge especially during time-dependent operation. Thus, research activities in the definition of surge precursors for control system inputs is deemed crucial [58].
- Optimization activities are needed to define the best configuration (e.g. installation of a recuperator or of a cathodic ejector), also considering the potential incorporation of low-temperature fuel cell technology.
- Low cost solutions are needed, including the exploitation of synergies between hybrid systems and turbocharger technology [22, 59].

7.7 Dynamic and Control System Aspects

Another important aspect to be considered to enable wide commercialization of hybrid systems involves the dynamic issues requiring special treatment in control system development. As stated in Chapter 4, the integration between two systems with such different time constants (high SOFC thermal capacitance and small turbine inertia) is critical to avoid risky conditions during dynamic operations. The importance of this aspect has generated interest from different research groups that developed various studies considering both theoretical [60–63] and experimental [64, 65] activities. As presented in Chapter 6, specific experimental facilities [64–66], named emulator rigs, were designed and installed to improve our engineering knowledge of the encountered dynamic aspects and to develop and test plant control systems without putting the most critical component (the SOFC stack) at risk of failure in case of malfunction. As mentioned previously, a number of results were presented using different control and plant solutions. The experimental tests based on these emulator rigs were focused on specific topics (e.g. surge prevention) and demonstrated positive solutions for SOFC/mGT matching in the sense of avoiding risk conditions (the details of which are presented in Chapter 4). However, due to a range of technical issues in developing prototypes, no real applications have been carried out so far, and there are no public data available from industrial applications.

Considering the state-of-the-art on dynamic and control system aspects, several research activities have to be carried out to achieve the maturity required at the commercial level. Specifically, the following aspects must be considered:

- Better assessment of the possible control approaches considering further calculations and tests (to demonstrate robustness and improve/validate risk prevention aspects);
- Integration of the results produced by researchers;
- Better coordination between the researchers involved in emulator-based tests, such as in [22];
- Collaborations between SOFC and turbine companies to perform tests on components and, finally, on hybrid system prototypes (this will be essential to assess the optimal control logic for use in hybrid systems).

7.8 CO_2 Separation Technologies for SOFC Hybrid Plants

Even though hybrid systems are considered to be highly efficient (due to the combined system efficiencies of the microturbine and fuel cells) and environmentally friendly technology (due to reduced emissions when using a fuel cell), they still include a considerable amount of CO_2 in the final exhaust stream if they are based on fossil fuels. Therefore, in order to reduce such emissions to a feasible minimum level, research activities on separation technologies [67–78] are of vital importance. This section presents some specific details of research activities for such approaches.

CO_2 separation based on the use of an aqueous solution of blended amines is studied in [71], with the focus on its operational behaviour and capital cost. The thermodynamic results show that despite a noticeable efficiency loss incurred due to fuel pre-treatment, CO_2 separation and compression, the results are still satisfactory in terms of efficiency.

CO_2 separation is one of a large number of alternative emission control solutions for power generation with hybrid systems [74, 75, 77]. Campanari [77] compared cycle configurations based on high-efficiency integrated SOFC-mGT hybrid cycles, where CO_2 is separated with absorption systems or with the eventual adoption of a second SOFC module acting as an 'afterburner' [77]. The author concluded that in addition to either a CO_2 separation technique applied to the cell anode exhaust or the adoption of a double fuel cell configuration, and when coupled with a corresponding gas turbine cycle, SOFCs can achieve fuel-to-electricity conversion efficiencies close to 70% with 90% CO_2 removal. This layout is also linked to additional advantages: there is no need for a high turbine inlet temperature, and there are near-zero NOx emissions. This efficiency level cannot be reached using any other conventional technology [78].

7.9 Coal and Biofuel for Hybrid Systems

The majority of hybrid systems developed to date have been built using either natural gas or hydrogen as a fuel [79–87]. Nevertheless, SOFC hybrid systems have huge advantages in terms of fuel flexibility. In this section, brief details covering research activities for configurations using coal and biofuel are presented. While a coal-based hybrid system could allow the utilization of a cheap fuel with an almost clean (with appropriate fuel treatment systems) and efficient plant, using biofuel in a hybrid system is considered one of the best solutions. The use of biofuel introduces the following important positive aspects in power generation for future applications: zero emissions (including CO_2), high efficiency, flexibility and continuous operations (avoiding the variability issues of renewable sources).

Lisbona & Romeo [87] analyzed an innovative concept that combines several of the essential future power generation subsystems: gasification, carbon looping, chemical looping, SOFC and CO_2 gas turbines. This paper also reported a global thermodynamic model relevant for fuel processing techniques to yield hydrogen and CO_2 capture methodologies combined with fuel and residual heat usage. Hydrogen was directly fed to a solid oxide fuel cell and exhaust streams were used in a gas turbine expander and in a heat recovery steam generator [87]. Also investigated was the influence of the steam to carbon ratio in the gasifier and the regeneration reactor, operating system pressure, optimum temperature ranges for oxygen transfer material oxidation, purge percentage in calcine, average sorbent activity and oxidant utilization in the fuel cell. The authors found that electrical efficiencies up to 73% are reached under optimal conditions, and the measured CO_2 capture efficiencies of near 96% ensure a good performance for greenhouse gas mitigation targets.

Another research effort based on the combination of coal gasification and a fuel cell for power generation was carried out by Chen *et al.* [78]. A thermodynamic analysis based on energy and exergy concepts was performed to investigate the performance of the integrated system and its sensitivity to the main operating parameters. It is well known that SOFCs produce electricity with high energy conversion efficiency, while chemical looping combustion (CLC) is employed as a process for fuel conversion with inherent CO_2 separation; hence, a novel combined cycle integrating coal gasification, solid oxide fuel cell operation, and chemical looping combustion was configured and analyzed. The simulations performed by the authors indicated that the plant net power efficiency reaches 49.8% with ≈100% CO_2 capture for an SOFC at 900°C, 15 bar, fuel utilization factor of 85%, and using NiO as the oxygen carrier in the CLC unit.

Sucipta *et al.* [84] performed an analysis of electricity generation efficiency for a biomass SOFC–mGT hybrid system, considering several cases covering different fuel compositions. They concluded that the efficiency metrics for all three cases of biomass fuel are reasonably high. SOFC–mGT hybrid systems show great promise when biomass-based fuels are used [84] with the aim of developing zero-emission, high-efficiency plants. However, in all three cases, the efficiency was lower than for the pure methane case, both in the SOFC module and in the hybrid system. Amongst the biomass fuel cases, efficiency is found to be the highest with steam-blown biomass fuel inputs both for the SOFC module and for the hybrid system.

7.10 Conclusions

Despite the great enthusiasm for hybrid system research and development that arose at the beginning of this new century, a slowdown of these activities is now experienced due to several reasons. At the beginning of the past decade, several authors forecast hybrid system commercialization in 5–7 years; however, technological, complexity and cost issues have resulted in a slowing-down of the initial research and development pace. Nevertheless, in comparison with the past decade, several publications have presented validated solutions for many earlier identified research issues, such as cost decreases, SOFC/mGT coupling and control system development. While the latest results are discussed in the previous chapters, several topics are still awaiting targeted and fundamental/applied research activities. Hence, a drive towards increasing the number of targeted studies and funding resources is still necessary to solve the technical issues that will, once solved, enable greater reliability, longer plant lifetime, lower costs and enhanced performance; these resources will help to support dedicated efforts in both academic and industrial sectors. Even though hybrid systems should not be expected to enter the market in the very near future, this technology is bound to be a central pillar for future energy generation and hydrogen economy development, as soon as all the problems discussed in this book are solved.

References

1 Thornton, A. and Monroy, C.R. (2011) Distributed power generation in the United States. *Renewable and Sustainable Energy Reviews*, 15, 4809–4817.

2 Pepermans, G., Driesen, J., Haeseldonckx, D., Belmans, R. and D'Haeseleer, W. (2005) Distributed generation: Definition, benefits and issues. *Energy Policy*, 33, 787–798.

3 Ferrari, M.L., Pascenti, M., Magistri, L. and Massardo, A.F. (2010) Micro gas turbine recuperator: Steady-state and transient experimental investigation. *Journal of Engineering for Gas Turbines and Power*, 132, 022301_1–8.

4 Fragaki, A., Andersen, A.N. and Toke, D. (2008) Exploration of economical sizing of gas engine and thermal store for combined heat and power plants in the UK. *Energy*, 33, 1659–1670.

5 Akorede, M.F., Hizam, H. and Pouresmaeil, E. (2010) Distributed energy resources and benefits to the environment. *Renewable and Sustainable Energy Reviews*, 14, 724–734.

6 Holladay, J.D., Jone, E.O., Phelps, M. and Hu, J. (2002) High-efficiency microscale power using a fuel processor and fuel cell. *Components and Applications for Industry, Automobiles, Aerospace, and Communication*, 4559, 148–156.

7 Ferrari, M.L. (2011) Solid oxide fuel cell hybrid system: Control strategy for stand-alone configurations. *Journal of Power Sources*, 196 (5), 2682–2690.

8 Liu, M., Lanzini, A., Halliop, W., Cobas, V.R.M., Verkooijen, A.H.M. and Aravind, P.V. (2013) Anode recirculation behavior of a solid oxide fuel cell system: A safety analysis and a performance optimization. *International Journal of Hydrogen Energy*, 38 (6), 2868–2883.

9 Hwang, J.J., Chang, W.R. and Su, A. (2008) Dynamic modeling of a solar hydrogen system under leakage conditions. *International Journal of Hydrogen Energy*, 33 (14), 3615–3624.

10 Loir, N. (2008) Energy resources and use: the present situation and possible paths to the future. *Energy*, 33 (6), 842–857.

11 Anvari-Moghaddam, A., Seifi, A., Niknam, T. and Reza Alizadeh-Pahlavani, M. (2011) Multi-objective operation management of a renewable MG (micro-grid) with back-up micro-turbine/fuel cell/battery hybrid power source. *Energy*, 36 (11), 6490–6507.

12 Ray, P.K., Mohanty, S.R. and Kishor, N. (2011) Proportional–integral controller based small-signal analysis of hybrid distributed generation system. *Energy Conversion and Management*, 52, 1943–1954.

13 Ferrari, M.L., Pascenti, M., Sorce, A., Traverso, A. and Massardo, A.F. (2014) Real-time tool for management of smart polygeneration grids including thermal energy storage. *Applied Energy*, 130, 670–678.

14 Bozzo, M., Caratozzolo, F. and Traverso, A. (2012) Smart polygeneration grid: Control and optimization system. ASME paper GT2012-68568, ASME Turbo Expo 2012, Copenhagen, Denmark.

15 Ferrari, M.L., Traverso, A., Pascenti, M. and Massardo, A.F. (2014) Plant management tools tested with a small-scale distributed generation laboratory. *Energy Conversion and Management*, 78, 105–113.

16 Ferrari, M.L., Rivarolo, M. and Massardo, A.F. (2016) Hydrogen production system from photovoltaic panels: Experimental characterization and size optimization. *Energy Conversion and Management*, 116, 194–202.

17 Loir, N. (2008) Energy resources and use: the present situation and possible paths to the future. *Energy*, 33 (6), 842–857.

18 US Department of Energy, Office of Fossil Energy. (2007) Vision 21 program. Available at http://www.fossil.energy.gov/programs/powersystems/vision21/index.html

19 Hwang, J.J., Chang, W.R. and Su, A. (2008) Dynamic modeling of a solar hydrogen system under leakage conditions. *International Journal of Hydrogen Energy*, 33 (14), 3615–3624.

20 Bellotti, D., Rivarolo, M., Magistri, L. and Massardo, A.F. (2015) Thermo-economic comparison of hydrogen and hydro-methane produced from hydroelectric energy for land transportation. *International Journal of Hydrogen Energy*, 40 (6), 2433–2444.

21 Leucht, F., Henke, M., Willich, C., Westner, C., Kallo, J. and Friedrich, K.A. (2012) Hydrogen rich natural gas as a fuel for SOFC systems. ICEPAG 2012, Costa Mesa, California, USA.

22 n.d http://www.bio-hypp.eu/

23 Erdinc, O. and Uzunoglu, M. (2012) Optimum design of hybrid renewable energy systems: Overview of different approaches. *Renewable and Sustainable Energy Reviews*, 16 (3), 1412–1425.

24 Yilmaz, P., Hocaoglu, M.H. and Konukman, A.E.S. (2008) A pre-feasibility case study on integrated resource planning including renewables. *Energy Policy*, 36 (3), 1223–1232.

25 Ferrari, M.L. (2011) Solid oxide fuel cell hybrid system: Control strategy for stand-alone configurations. *Journal of Power Sources*, 196 (5), 2682–2690.

26 Mellit, A., Benghanem, M. and Kalogirou, S.A. (2007) Modeling and simulation of a standalone photovoltaic system using an adaptive artificial neural network: Proposition for a new sizing procedure. *Renewable Energy*, 32 (2), 285–313.

27 Anagnostopoulos, J.S. and Papantonis, D.E. (2007) Pumping station design for a pumped storage wind hydro power plant. *Energy Conversion and Management*, 48 (11), 3009–3017.

28 Phuangpornpitak, N. and Kumar, S. (2007) PV hybrid systems for rural electrification in Thailand. *Renewable and Sustainable Energy Reviews*, 11 (7), 1530–1543.

29 Franzoni, A., Magistri, L., Traverso, A. and Massardo, A.F. (2006) Thermoeconomic analysis of hybrid SOFC fuel cell systems with decarbonization of natural gas. ASME paper GT2006-90562, ASME Turbo Expo 2006, Barcelona, Spain.

30 Ter-Gazarian, A.G. and Kagan, N. (1992) Design model for electrical distribution systems considering renewable, conventional and energy storage units. *IEE Proceedings C: Generation, Transmission and Distribution*, 139 (6), 499–504.

31 Anglani, N. and Muliere, G. (2010) Analyzing the impact of renewable energy technologies by means of optimal energy planning. 9th International Conference on Environment and Electrical Engineering (EEEIC), pp. 1–5.

32 Mizani, S. and Yazdani, A. (2009) Design and operation of a remote microgrid. 35th Annual Conference on Industrial Electronics (IECON'09), pp. 4299–4304.

33 HOMER Energy LLC. (2015) Available at http://www.homerenergy.com/company.html

34 Isherwood, W., Smith, J.R., Aceves, S.M., Berry, G., Clark, W., Johnson, R., Das, D., Goering, D. and Seifert (2000)

Remote power systems with advanced storage technologies for Alaskan villages. *Energy*, 25 (10), 1005–1020.

35 Darras, C., Sailler, S., Thibault, C., Muselli, M., Poggi, P., Hoguet, J.C., Melscoet, S., Pinton, E., Grehant, S., Gailly, F., Turpin, C., Astier, S. and Fontès, G. (2010) Sizing of photovoltaic system coupled with hydrogen/oxygen storage based on the ORIENTE model. *International Journal of Hydrogen Energy*, 35 (8), 3322–3332.

36 n.d http://www.tpg.unige.it/TPG/

37 Porta, M., Traverso, A. and Marigo, L. (2006) Thermoeconomic analysis of a small-size biomass gasification plant for combined heat and distributed power generation. ASME Paper No. GT2006-90918, ASME Turbo Expo, Barcelona, Spain.

38 Efren, O. and Efren, B.Y. (2010) Size optimization of a PV/wind hybrid energy conversion system with battery storage using simulated annealing. *Applied Energy*, 87 (2), 592–598.

39 Nema, P., Nema, R.K. and Rangnekar, S. (2009) A current and future state of art development of hybrid energy system using wind and PV-solar: A review. *Renewable and Sustainable Energy Reviews*, 13 (8), 2096–2103.

40 Connolly, D., Lund, H., Mathiesen, B.V. and Leahy, M. (2010) A review of computer tools for analysing the integration of renewable energy into various energy systems. *Applied Energy*, 87 (4), 1059–1082.

41 Nehrir, M.H. and Wang, C. (2009) *Modelling and Control of Fuel Cells: Distributed Generation Applications*. Vol. 41. John Wiley & Sons, Chichester.

42 Costamagna, P., Magistri, L. and Massardo, A.F. (2000) Techno economical analysis of SOFC reactor gas turbine combined plants. IEA Meeting, Modeling and Simulation of Hybrid SOFC-GT Systems and Components, Ystad.

43 Massardo, A.F. and Magistri, L. (2001) Internal Reforming Solid Oxide Fuel Cell Gas Turbine Combined Cycles (IRSOFC-GT). Part II: Exergy and thermoeconomic analyses. *Journal of Engineering for Gas Turbines and Power*, 125, 67–74.

44 Wilson, J.D., Colson, C.M. and Nehrir, M.H. (2010) Cost and unit-sizing analysis of a hybrid SOFC/microturbine

generation system for residential applications. IEEE North American Power Symposium (NAPS), 2010.

45 Wang, C. and Nehrir, M.H. (2007) A physically based dynamic model for solid oxide fuel cells. *IEEE Transactions on Energy Conversion*, 22, 887–897.

46 Agnew, G.D., Bozzolo, M., Moritz, R.R. and Berenyi, S. (2005) The design and integration of the Rolls-Royce Fuel Cell Systems 1MW SOFC. ASME Paper GT2005-69122.

47 Haga, K., Adachi, S., Shiratori, Y., Itoh, K. and Sasaki, K. (2008) Poisoning of SOFC anodes by various fuel impurities. *Solid State Ionics*, 179, 1427–1431.

48 Park, K., Yu, S., Bae, J., Kim, H. and Ko, Y. (2010) Fast performance degradation of sofc caused by cathode delamination in long-term testing. *International Journal of Hydrogen Energy*, 35, 8670–8677.

49 Vladikova, D.E., Stoynov, Z.B., Barbucci, A. and Viviani, M. (2008) Impedance studies of cathode/electrolyte behavior in SOFC. *Electrochimica Acta*, 53, 7491–7499.

50 Offer, G.J. and Brandon, N.P. (2009) The effect of current density and temperature on the degradation of nickel cermet electrodes by carbon monoxide in solid oxide fuel cells. *Chemical Engineering Science*, 64, 2291–2300.

51 Zaccaria, V., Tucker, D. and Traverso, A. (2016) A distributed real-time model of degradation in a solid oxide fuel cell, part I: Model characterization. *Journal of Power Sources*, 311, 175–181.

52 Milewski, A. and Miller, J. (2011) SOFC-GT hybrid system – turbomachinery problems. 9th European Conference on Turbomachinery – Fluid Dynamics and Thermodynamics.

53 Guleren, K.M., Afgan, I. and Turan, A. (2010) Predictions of turbulent flow for the impeller of a NASA low-speed centrifugal compressor. *Journal of Turbomachinery*, 132 (2), 021005_1–8.

54 Greco, A., Sorce, A., Littwin, R., Costamagna, P. and Magistri, L. (2014) Reformer faults in SOFC systems: Experimental and modeling analysis, and simulated fault maps. *International Journal of Hydrogen Energy*, 39 (36), 21700–21713.

55 Barber-Nichols Inc. (2013) Guidelines for Determining Foil Bearing Applicability, Arvada. Available at www.barber-nichols.com

56 Staunton, R.H. and Ozpineci, B. (2003) *Microturbine Power Conversion Technology Review*. Report ORNL/TM-2003/74, Oak Ridge National Laboratory.

57 McDonald, C.F. (2000) Low-cost compact primary surface recuperator concept for microturbines. *Applied Thermal Engineering*, 20, 471–497.

58 Aeristech Limited and Matthew Houldershaw (2017) efficient, low cost turbo-machine compressor for automotive fuel cell systems. TSB Project 101580, Aeristech. Available at http://gtr.rcuk.ac.uk/projects?ref=101580

59 Damo, U.M., Ferrari, M.L., Turan, A. and Massardo, A.F. (2015) Re-compression model for SOFC hybrid systems: Start-up and shutdown test for an emulator rig. *Fuel Cells*, 1, 42–48.

60 Ferrari, M.L. (2015) Advanced control approach for hybrid systems based on solid oxide fuel cells. *Applied Energy*, 145, 364–373.

61 Nanaeda, K., Mueller, F., Brouwer, J. and Samuelsen, S. (2010) Dynamic modeling and evaluation of solid oxide fuel cell – combined heat and power system operating strategies. *Journal of Power Sources*, 195, 3176–3185.

62 Santis-Alvarez, A.J., Nabavi, M., Hild, N., Poulikakos, D. and Stark, W.J. (2011) A fast hybrid start-up process for thermally self-sustained catalytic n-butane reforming in micro-SOFC power plants. *Energy and Environmental Science*, 4, 3041–3050.

63 McLarty, D., Brouwer, J. and Samuelsen, S. (2014) Fuel cell gas turbine hybrid system design part II: Dynamics and control. *Journal of Power Sources*, 254, 126–136.

64 Tucker, D., Lawson, L., VanOsdol, J., Kislear, J. and Akinbobuyi, A. (2006) Examination of ambient pressure effects on hybrid solid oxide fuel cell turbine system operation using hardware simulation. ASME Paper No. 2006-GT-91291.

65 Ferrari, M.L., Pascenti, M., Magistri, L. and Massardo, A.F. (2010) Hybrid system test rig: Start-up and shutdown physical emulation. *Journal of Fuel Cell Science and Technology*, 7, 021005_1–7.

66 Hohloch, M., Widenhorn, A., Lebküchner, D., Panne, T. and Aigner, M. (2008) Micro gas turbine test rig for hybrid power plant application. ASME Paper GT2008-50443.

67 Zabihian, F. and Fung, A. (2009) A review on modelling of hybrid solid oxide fuel cell systems. *International Journal of Engineering*, 3, 85–119.

68 Riensche, E., Achenbach, E., Froning, D., Haines, M.R., Heidug, W.K., Lokurlu, A. and Andrian, S. von (2000) Clean combined-cycle SOFC power plant – cell modelling and process analysis. *Journal of Power Sources*, 86, 404–410.

69 Riensche, E., Achenbach, E., Froning, D., Haines, M.R., Heidug, W.K., Lokurlu, A. and Von Andrian, S. (2000) Clean combined-cycle SOFC power plant – cell modelling and process analysis. *Journal of Power Sources*, 86 (1), 404–410.

70 Achenbach, E. (1994) Three-dimensional and time-dependent simulation of a planar solid oxide fuel cell stack. *Journal of Power Sources*, 49, 333–348.

71 Franzoni, A., Magistri, L., Traverso, A. and Massardo, A.F. (2008) Thermoeconomic analysis of pressurized hybrid SOFC systems with CO_2 separation. *Energy*, 33, 311–320.

72 Massardo, A.F. and Lubelli, F. (2000) Internal reforming solid oxide fuel cell-gas turbine combined cycles (IRSOFC-GT): Part A – Cell model and cycle thermodynamic analysis. *Journal of Engineering for Gas Turbines and Power*, 122, 27–35.

73 Inui, Y., Yanagisawa, S. and Ishida, T. (2003) Proposal of high performance SOFC combined power generation system with carbon dioxide recovery. *Energy Conversion and Management*, 44, 597–609.

74 Campanari, S. and Chiesa, P. (2002) Potential of solid oxide fuel cells (SOFC) based cycles in low-CO2 emission power generation. Proceedings of the 6th International Conference on Greenhouse Gas Control Technologies, Kyoto, Japan.

75 Campanari, S. (2001) Thermodynamic model and parametric analysis of a tubular SOFC module. *Journal of Power Sources*, 92, 26–34.

76 Kuramochi, T., Wu, H., Ramírez, A., Faaij, A. and Turkenburg, W. (2009) Techno-economic prospects for CO_2 capture from a solid oxide fuel cell–combined heat and power plant. Preliminary results. *Energy Procedia*, 1 (1), 3843–3850.

77 Campanari, S. (2002) Carbon dioxide separation from high temperature fuel cell power plants. *Journal of Power Sources*, 112 (1), 273–289.

78 Chen, S., Lior, N. and Xiang, W. (2015) Coal gasification integration with solid oxide fuel cell and chemical looping combustion for high-efficiency power generation with inherent CO_2 capture. *Applied Energy*, 146, 298–312.

79 European Environment Agency (2005) Annual European Community greenhouse gas inventory 1990–2003 and inventory report 2005. Technical Report No. 4. European Environment Agency, Copenhagen.

80 Lobachyov, K. and Richter, H.J. (1996) Combined cycle gas turbine power plant with coal gasification and solid oxide fuel cell. *Journal of Energy Resources Technology*, 118, 285–292.

81 Kivisaari, T., Björnbom, P., Sylwan, C., Jacquinot, B., Jansen, D. and de Groot, A. (2004) The feasibility of a coal gasifier combined with a high-temperature fuel cell. *Chemical Engineering Journal*, 100, 167–180.

82 Kuchonthara, P., Bhattacharya, S. and Tsutsumi, A. (2005) Combination of thermochemical recuperative coal gasification cycle and fuel cell for power generation. *Fuel*, 84, 1019–1021.

83 Rao, A.D., Verma, A. and Samuelsen, G.S. (2005) Engineering and economic analyses of a coal-fueled solid oxide fuel cell hybrid power plant. Proceedings of the 2005 ASME Turbo Expo., Reno-Tahoe, US.

84 Sucipta, M., Kimijima, S. and Suzuki, K. (2007) Performance analysis of the SOFC–MGT hybrid system with gasified biomass fuel. *Journal of Power Sources*, 174, 124–135.

85 Van Herle, J., Maréchal, F., Leuenberger, S. and Favrat, D. (2003) Energy balance model of a SOFC cogenerator operated with biogas. *Journal of Power Sources*, 118, 375–383.

86 Raak, H., Diethelm, R. and Riggenbach, S. (2002) The Sulzer Hexis story: from demonstrators to commercial products. Proceedings of the Fuel Cell World, Lucerne, Switzerland.

87 Lisbona, P. and Romeo, L.M. (2008) Enhanced coal gasification heated by unmixed combustion integrated with a hybrid system of SOFC/GT. *International Journal of Hydrogen Energy*, 33 (20), 5755–5764.

Glossary

Glossary	Symbol/ Acronym	Definition
Fuel cell		Device converting the chemical energy of a fuel into electrical work through electrochemical oxidation
Hybrid system		Power generation system resulting from the thermal integration of a high-temperature fuel cell and a micro gas turbine
Microturbines	mGTs	Small heat engine whose operating principle relies on the Brayton-Joule cycle
Conventional energy sources		Non-renewable energy sources employed to cover the largest share of the electricity demand: mostly fossil and nuclear fuels
Global warming		Long-term global temperature rise as a a consequence of greenhouse gas emissions
Fossil fuel		Combustible geological deposit formed by the anaerobic decomposition of organic material
Renewable energy sources		Energy resources which can be naturally replenished on a human timescale
Energy density		Energy content per unit of volume
Intermittency		Non-constant supply of renewable energy
Greenhouse gases		Gases (usually present in the atmosphere) that absorb and emit radiation in infrared wavelength range
Power generation		Conversion of primary energy into electricity that can be used for human needs

Hybrid Systems Based on Solid Oxide Fuel Cells: Modelling and Design, First Edition.
Mario L. Ferrari, Usman M. Damo, Ali Turan, and David Sánchez.
© 2017 John Wiley & Sons Ltd. Published 2017 by John Wiley & Sons Ltd.

Glossary	Symbol/ Acronym	Definition
Energy efficiency		Relation between the energy supplied to a system and the work developed by such system when performing an energy conversion process
CO_2 emissions		Emissions of carbon dioxide, usually referred to power generation systems
Thermal power plants		Facilities that produce electricity relying on thermo-mechanical energy conversion processes
Working cycle		Thermodynamic process aimed at converting heat into useful work cyclically
Combined heat and power	CHP	Power generation system that can take advantage of residual thermal energy, usually wasted in conventional power plants
First law efficiency		Fraction/percentage of the total heat energy supplied to a system that is converted to net work
Primary energy		Energy embodied in sources without prior separation or cleaning (i.e. only extraction or capture) – in this context, raw energy supplied to power generation systems
Biomass		Renewable energy resource formed by combustible organic matter that is either living or living until recently (as opposed to fossil fuels)
Solar power		Renewable energy resource from solar radiation captured at ground level
Wind farms		Power station made up of a large number of wind turbines
Biofuel		Fuel produced from organic matter through contemporary biological processes, i.e. not in a self-driven natural process on a century-long timescale
Electrical efficiency		In a power generation process, ratio of the electric power produced to the primary energy supply for a power generation process
Solid oxide fuel cell	SOFC	High-temperature fuel cell with solid state electrolyte
Anode		Electrode with negative charge (fuel electrode)

Glossary	Symbol/ Acronym	Definition
Cathode		Electrode with positive charge (air electrode)
Electrode		Porous element of a fuel cell where the oxidation and reduction chemical reactions take place
Electrolyte		Element of a fuel cell across which charge in the form of ions is transferred between electrodes
Electrolysis		Process whereby water molecules are split into oxygen and hydrogen
Internal energy	U	Energy associated with the motion of microscopic particles
Helmholtz free energy	A	Work potential that represents the maximum work that can be performed by a system in a process at constant volume and temperature
Enthalpy	H	Energy content of an open system resulting from the addition of internal energy and flow work
Gibbs free energy	G	Work potential that represents the maximum work that can be performed by a system in a process at constant pressure and temperature
Entropy	S	Measure of the microscopic randomness of a closed system
Irreversible process		A change in the thermodynamic state of a system and its surroundings which cannot be restored without additional expenditure of energy
Isentropic process		Change in the thermodynamic state of a system that takes place at constant entropy. Also, reversible and adiabatic thermodynamic process
Isothermal process		Change in the thermodynamic state of a system that takes place at constant temperature
Alkaline fuel cell		A fuel cell whose electrolyte is made up of a porous matrix saturated with an alkaline solution and operating with hydrogen and oxygen as fuel and oxidant respectively

Glossary	Symbol/ Acronym	Definition
Proton exchange membrane	PEM	A polymerous electrolyte that conducts protons (positive ions) from anode to cathode
Electrical conductivity		Ability (or capacity) to conduct electrical charge (whether electrons or ions)
Zirconia cermet		A composite material made of metal (usually nickel) and zirconium dioxide
Sintering		Process where metallic powders form a congruent solid by the application of heat or pressure, without achieving the melting point of the substances (point of liquefaction)
Coefficient of thermal expansion		Ratio between the elongation and the temperature rise of a piece of solid material
Porosity		Fraction of empty volume (voids) with respect to total volume
Perovskite structures		Crystalline structure of the form ABX_3, where A is in the cube corners, B in the centre and X (usually oxygen) occupies the center of the faces
Interconnector		Connection between two or more fuell cells operating in parallel.
Poisoning		Deactivation of a catalyst due to interaction with certain species in the fuel or air streams
Fuel reforming		Chemical process whereby a hydrogen-rich syngas is obtained from a hydrocarbon fuel
Steam reforming		Reforming of a hydrocarbon fuel mixed with steam at high temperature
Partial oxidation		Hydrocarbon fuel reforming based on partial combustion with air to produce a hydrogen-rich gas
Autothermal reforming	ATR	Reforming process based on the partial combustion of a hydrocarbon fuel with oxygen and carbon dioxide or steam to produce a hydrogen-rich gas

Glossary	Symbol/ Acronym	Definition
Gasification		Process whereby a hydrocarbon fuel reacts with oxygen and steam at high temperature and pressure to produce a hydrogen-rich gas
Standard temperature and pressure	STP	Standard values defined by the International Union of Pure and Applied Chemistry: 298.15 K and 100 kPa
Faraday's law		Quantitative relation between the moles of hydrogen consumed and the electric current established between electrodes in a fuel cell
Faraday's constant	F	Electric charge of a mole of electrons
Nernst potential	E	Voltage difference between electrodes in an ideal, reversible fuel cell with no losses
Le Chatelier's principle		Principle stating that a system in equilibrium subjected to a change in concentration, pressure, volume or temperature will react so as to counteract the said change
Dalton's law		Law stating that the total pressure exerted by a mixture of non-reacting gases equals the sum of the partial pressures of the individual gases
Activation losses		Voltage drop needed to overcome the activation energy of the chemical reactions taking place in the fuel cell
Fuel crossover		Fuel diffusion through the electrolyte which results in a voltage loss
Ohmic losses		Voltage losses caused by the ionic resistance of electrolyte and electrodes, and the electronic resistance of the electrodes, collectors and interconnects
Mass diffusion loss		Loss due to the limited diffusion of reactants and products and the consequent reduction in reaction rates in a fuel cell
Ohm's law		Law stating that the current intensity between two points of an electric circuit is proportional to the voltage drop between the said locations

Glossary	Symbol/ Acronym	Definition
Portable power generation		Power generation system that is easy to transport or to embed in a vehicle
Stationary power generation		Non-portable power generation system
Recuperated Brayton cycle		Brayton cycle where the waste heat downstream of the turbine is used to preheat the combustion air delivered by the compressor
Carnot cycle		Thermodynamic cycle comprising two processes at constant temperature and two processes at constant entropy. It sets the maximum efficiency attainable by a cycle converting heat into mechanical work when operating between two reservoirs at different temperatures
Molten carbonate fuel cell	MCFC	High-temperature fuel cell whose electrolyte is made up of molten carbonate salts of lithium, sodium and potassium stored in a porous lithium aluminate matrix
Inverter		An electronic device converting direct current into alternating current (DC-AC)
Noble-metal based catalyst		Catalyst used in the electrodes of a fuel cell to promote the reduction/oxidation of reactants (typically platinum)
Reformed gas		Hydrogen-rich syngas resulting from hydrocarbon fuel reforming
Fuel cleaning device		Device to filter out the impurities of a fuel by physical and/or chemical means
Bottoming cycle		A cycle employed for further waste heat recovery in a thermal process, for instance the hot flue gases in a thermal power generation system
Cogenerative applications		Facility producing heat and electric power at the same time. See Combined Heat and Power also
Superheated steam		Steam at a temperature above the corresponding saturation temperature

Glossary	Symbol/ Acronym	Definition
Stack		Referred to fuel cells, series-parallel array of elementary fuel cells aimed at achieving higher voltage and current (i.e. power)
Yttria-stabilized zirconia	YSZ	Ceramic material made of a crystal structure of zirconium dioxide doped with yttrium oxide
Ferritic stainless-steels		Stainless steel with low chromium and nickel content
Chromium based material		Alloy with high chromium content
Doped ceria		Cerium dioxide doped with, typically, gadolinium in order to increase the ionic conductivity at lower temperatures
Lanthanum gallate		Lanthanum-based material comprising also strontium, gallium and magnesium oxides with high ionic conductivity to be used in high-temperature fuel cell electrolytes
Power density		Power output per unit active (usually electrode) area of a fuel cell
Triple phase boundary		Reaction site where the gas phase comes into contact with the electrode and electrolyte simultaneously
Reaction kinetics		Factors controlling the rate at which a chemical reaction proceeds
Scandium-doped zirconia	SDZ	Zirconium dioxide doped with scandium for enhanced conductivity
Catalytic activity		Capacity of a given catalyst material to increase the rate at which a particular reaction proceeds (by reducing the corresponding activation energy)
Ionic conductor		Material that enables the transportation of ions across its internal structure
Thermal cycling		Processes whereby a certain specimen or system is subjected to cyclic changes in temperature due to frequent changes in the operating conditions

Glossary	Symbol/Acronym	Definition
Tubular SOFCs		Solid oxide fuel cells whose stack has a tubular geometry
Strontium-doped lanthanum manganite	LSM	Ceramic material formed by manganese and lanthanum oxides and doped with strontium for enhanced electronic conductivity, typically used in SOFC cathodes
Co-flow fuel cell		Tubular SOFC where the oxidant and fuel streams flow in parallel (from the bottom to the top)
Off-gas burner	OGB	Combustor where the excess fuel (from the anode) of a fuel cell is burnt
Slurry deposition		Superficial treatment wherein a covering material is provided in the form of a liquid dissolution
Plasma spray		Superficial treatment in which the covering material is projected over the base at high temperature, high energy and high velocity so the gas becomes ionized (plasma)
Segmented-in-series tubular SOFCs		Tubular SOFCs connected in series in such a way that the geometric complexity and manufacturing costs are reduced
IP-SOFC		Segmented-in-series tubular SOFC on a flattened tube configuration
Planar SOFC		Fuel cell formed by flat plates connected in series
Self-supporting layout		Fuel cell configuration in which the electrodes and/or the electrolyte provide mechanical resistance (structural function)
External-supporting layout		Fuel cell configuration in which the structural support is provided by an external element
Fuel processing		Process whereby a hydrogen-rich syngas is obtained from a raw (usually hydrocarbon) fuel
Pre-reformer		Small fuel reactor located upstream of the fuel cell, which initiates the fuel reforming process

Glossary	Symbol/ Acronym	Definition
Indirect internal reforming	IIR	Reforming process in which there is heat exchange between the reformer and the fuel cell stack
Direct internal reforming	DIR	Reforming process in which there is heat and mass exchange between the reformer and the fuel cell stack
Substoichio-metric mixture		Lean fuel and air mixture (the amount of fuel is lower than stoichiometric)
Partial combustion		Combustion reaction in which the fuel is not completely consumed
Moving bed		Chemical reactor in which the reactants and catalyst flow through the reactor
Fluidized bed		Chemical reactor in which solid particles (usually catalyst) are suspended in a gaseous flow (reactants) so that the former acquires a fluid-type behaviour
Entrained bed		Chemical reactor in which small particles of a solid fuel are injected along with the oxidant gas (air or oxygen), thus favouring the intimate contact between reactants
Water-gas shifting		Chemical process whereby carbon monoxide is oxidized to carbon dioxide in the presence of steam, thus releasing hydrogen molecules
Desulfurizer reactors		Chemical reactor which reduces the sulfur dioxide content of a gaseous stream (whether gas fuel or flue gases)
Cyclones		Mechanical device designed to remove solid particles from a gaseous stream by means of centrifugal effects
Electrostatic precipitators		Filtration device designed to capture small particles by means of electrostatic forces
Atmospheric SOFC hybrid systems		Combined SOFC and bottoming heat engine in which the fuel cell operates at atmospheric pressure
Recirculation		Reutilization of the fuel cell outflow into the feed stream to fully exploit the fuel content

Glossary	Symbol/ Acronym	Definition
Pressurized SOFC hybrid systems		Combined SOFC and bottoming heat engine in which the fuel cell is pressurized by the compressor in the bottoming gas turbine
Capital costs		All the one-time expenses needed to put a project into operation
Delamination		Failure of laminated materials in which internal layers separate from each other
Migration		Transportation of ions in the presence of a motive force
Degradation rate		Percentage of voltage drop per hour
Fuel utilization factor		Fraction of the fuel that is oxidized electrochemically in a fuel cell (i.e. excluding oxidation in the off-gas burner)
Isobaric specific heat	$\bar{c}_p$	Slope of the enthalpy versus temperature curve for a substance subjected to a change in thermodynamic state at constant pressure
Useful specific work	W_u	Useful work performed by a system per unit mass of circulating working fluid
Efficiency		
thermal	η_B	Efficiency of a thermodynamic cycle excluding any external irreversibility (based on the First Law of Thermodynamics, ratio between useful specific work and heat input to the system)
isentropic	η_c, η_t	Ratio of isentropic enthalpy change and actual enthalpy change across a turbomachinery (referred to the largest of the two so as to limit it to 100%)
Carnot	η_C	Efficiency of a Carnot cycle working between the reference temperatures of the hot and cold reservoirs
heat transfer	R	Ratio of actual heat exchange to the maximum heat exchange that is possible in a given heat exchanger (minimum heat capacity times maximum temperature change)
total-to-total	η_{tt}	Isentropic efficiency of a turbomachinery not accounting for kinetic energy losses

Glossary	Symbol/ Acronym	Definition
total-to-static	η_{ts}	Isentropic efficiency of a turbomachinery accounting for kinetic energy losses
Polytrophic process		General thermodynamic process according to the law $p \cdot V^n = C$ where C is constant
Perfect gas		Ideal gas with constant isobaric specific heat and specific heat ratio
Specific heat ratio	γ	Ratio from isobaric specific heat to isochoric specific heat. Also, isentropic exponent for a perfect gas
Pseudo-pressure ratio	δ	Temperature ratio along an isentropic process, which can be described as a function of pressure ratio and ratio of specific heats
Temperature ratio	θ	Ratio of turbine inlet temperature to compressor inlet temperature in a Brayton cycle
Quasi-ideal Brayton cycle		Brayton cycle working with an ideal gas at constant mass flow rate and composition across the entire cycle, with no pressure losses and adiabatic irreversible compression and expansion processes
Recuperated Brayton cycle		Brayton cycle where the waste heat downstream of the turbine is used to preheat the combustion air delivered by the compressor
Recuperator		Heat exchanger transferring thermal energy (heat) from turbine exhaust to compressor outlet
Intercooling		Heat rejection between compression stages
Reheating		Heat addition between turbine stages
Compound cycle		Thermodynamic cycle including intercooling, reheating or both
Microturbines (micro gas turbines)		Small gas turbines, with outputs below 500 kWe, normally used in distributed generation applications
Stage head coefficient	ψ_s	Pseudo non-dimensional isentropic enthalpy change across a turbomachinery stage
Stage loading coefficient	ψ	Pseudo non-dimensional enthalpy change across a turbomachinery stage

Glossary	Symbol/ Acronym	Definition
Flow coefficient	ϕ	Ratio of meridional flow speed to peripheral speed
Meridional flow speed	c_m	Projection of the flow velocity vector onto the meridional plane
Velocity ratio		Ratio of peripheral speed to isentropic velocity
Peripheral velocity	u	Blade speed at a given radius
Isentropic velocity	c_s	Theoretical flow velocity of the outlet of an isentropic nozzle, given the total inlet pressure and temperature and the static backpressure
Degree of reaction	R	Ratio from static enthalpy change across the rotor to total enthalpy change across a turbomachinery stage
Specific speed	ω_s	Non-dimensional parameter indexing impeller rotating speed for given volumetric flow rate and head
Specific diameter	D_s	Non-dimensional parameter indexing impeller size for given volumetric flow rate and head
Mach number	M	Ratio of flow velocity to speed of sound
Reynolds number	Re	Ratio of inertial forces to viscous forces in a fluid flow
Cordier line		Loci of specific speed and diameter for peak turbomachinery efficiency
Euler equation		Fundamental equation relating velocity diagrams at impeller inlet and outlet (mainly whirl velocity change) to stage-specific work
Whirl velocity	c_θ	Tangential velocity
Radii ratio		Ratio of largest to smallest radii in the rotor of a radial turbomachinery stage
Guide vanes		Vanes upstream of the compressor impeller that impart a certain whirl velocity distribution to the inlet flow
Inducer		Entry region of the impeller responsible for turning the flow from the axial to the radial direction

Glossary	Symbol/ Acronym	Definition
Sweep angle	β	Relative flow angle at the outlet of a radial compressor impeller
Radial blades		Purely radial (straight) blades in a centrifugal compressor impeller
Diffuser		Stator in a radial compressor
Backswept blades		Backward curved blades in a radial compressor impeller
Friction losses		Total pressure loss due to the effect of fluid viscosity near the walls
Height ratio		Rotor height ratio between inlet and outlet
Vaned diffuser		Stator of a centrifugal compressor with blades forming a divergent gas path
Vaneless diffuser		Stator of a centrifugal compressor without blades
Incidence losses		Aerodynamic losses due to misalignment between blade and flow at the leading edge of a blade row in a turbomachinery
Blockage losses	α	Aerodynamic losses brought about by the limited flow capacity in a turbomachinery
Wedge diffuser		Radial compressor diffuser with wedge-shaped vanes
Compressor surge		Generalized instability of a compressor resulting in flow reversal and inoperability
Stall		Local flow detachment in the gas path of a turbomachinery
Surge line		Loci of the minimum corrected mass flow rate ensuring stable operation for a given corrected rotational speed of a turbocompressor
Running line		Loci of the actual operating conditions of a turbocompressor for variable corrected speed
Surge margin	*SM*	Distance between the running and surge lines
Discharge (blow-off) valve		Fast-acting valve located downstream of the compressor to relieve pressure in case of a sudden drop in mass flow

Glossary	Symbol/ Acronym	Definition
Recycle valve		Valve located downstream of the compressor to increase the throughput by recirculation of a fraction of the outlet flow
Variable geometry		Stationary elements of a turbomachinery (blades or vanes) which can change their stagger angle depending on the flow (operating) conditions
Swirl		Tangential velocity
Clearance losses		Total pressure loss due to the leakage flow through the interstitial gaps between stationary and rotating elements along the gas path of a turbomachinery
Primary surface heat exchangers		Heat exchanger without extended heat exchange surfaces
Regenerator matrix		Rotating part of a regenerator that is alternately put into contact with the hot and cold flows
Conductance ratio	C_r	Ratio between the products of heat transfer coefficients and heat exchange areas in the hot and cold sides of a recuperator
Magnetic bearings		Bearings that make use of electromagnetic forces to levitate the shaft, in order to reduce friction losses
Conformal fluid film bearings		Bearings where the shaft is supported on a thin fluid lubricant film
Hydrodynamic bearing		Bearing in which the lubricant pressure is provided by the motion of the shaft itself
Hydrostatic bearing		Bearing in which the lubricant pressure is provided by an external device
Gas bearings		Conformal fluid film bearings using a lubricant in the gas phase
Tilting pad		Bearing in which a number of pads are inserted in the gap between the rotating shaft and the journal support in order to avoid misalignment and reduce losses
Sommerfeld number	S	Ratio of hydrodynamic pressure to specific load per unit area in a bearing
Distributed power generation		Decentralized energy (power or combined heat and power) generation by a variety of small-scale devices

Glossary	Symbol/ Acronym	Definition
Combined cycle		Combination of thermodynamic power cycles that exchange heat between one another, thereby increasing the overall energy conversion efficiency
Electrical efficiency		Ratio from electric output to primary energy input in a power generation system
Heat recovery steam generator	HRSG	Heat exchanger that produces steam from waste heat that is typically available at intermediate temperature
Blade cooling		Convective cooling system incorporated in high-pressure turbines to enable high firing temperatures at moderate metal temperatures (below melting point)
PID controllers	PID	Control loop feedback mechanism that calculates an error between a measurement and a target value, and applies a correction based on proportional, integral and derivative terms
Feed-forward		Control method in which the reaction to a system disturbance is based on previous knowledge of the system's expected behaviour, not on the actual system response
Model predictive control		Control method based on the predictions of a mathematical/analytical model of the system
H-infinity		Control method to synthesize controllers for multivariate systems, ensuring stability and performance
CO_2 sequestration system		Process whereby CO_2 from hydrocarbon fuel oxidation is separated from a gaseous flow (typically a stream of exhaust gases in a power generation system)
Stand-alone mode		Power system that operates without connection to the grid (off-grid). It might also refer to power systems that are not combined with others to yield higher global efficiency
Steam-to-carbon ratio		Molar ratio between the water steam and fuel carbon contents at the inlet of a steam reformer

Glossary	Symbol/ Acronym	Definition
2D approaches		Two-dimensional models
1D approaches		One-dimensional models
0D approaches		Zero-dimensional models
Real-time modelling approaches		Model that can be run/solved at the same rate as the actual physical process being simulated
Plant monitoring		Combination of hardware and software that provides the operator with data about the actual state of the power plant
Fault detection		The ability to (analytically or experimentally) detect the abnormal operation of a component in a system
Model validation		Verification that the model predictions match the experimental data available so as to accept the accuracy of the model
Look-up table		Matrix whose data are later interpolated by the solver during model resolution
Black box modelling		see Lumped volume
Single-shaft unit		Gas turbine where compressor, turbine and electric generator are mounted on the same (single) shaft
Free turbine engine		Gas turbine where the compressor and high-pressure turbine are mounted on one (high pressure) shaft whilst the low-pressure turbine and electric generator are mounted on a different (low pressure) shaft
Mass flow function		Relation between mass flow, pressure and specific volume at turbine inlet, used to model off-design performance of the turbine
Gas generator		In a free turbine engine, the compressor, combustor (heat adder) and high-pressure expander producing hot gas for the power (low pressure) turbine
Free power turbine		In a free turbine engine, the turbine driving the load
Reversible cell voltage	E	see Nernst potential

Glossary	Symbol/ Acronym	Definition
Overpotential	V_{loss}	Voltage loss in a fuel cell
Overpotential at the anode	η_a	Difference between the thermodynamic potential of the half reaction in the anode and the potential experimentally observed in the redox reaction
Overpotential at the cathode	η_c	Difference between the thermodynamic potential of the half reaction in the cathode and the potential experimentally observed in the redox reaction
Ohmic losses	η_{ohmic}	Voltage drop due to the electrical/ionic resistance of the fuel cell elements (electrolyte and collector plates mainly)
Air utilization factor	U_A	Fraction of oxygen (in air) that is effectively oxidized in the cathode of a fuel cell
Lumped volume		Modelling technique based on reducing a system to a number of discrete volumes or "lumps", within which properties are assumed constant
Emulator test rigs		Experimental facility where a subsystem/component is substituted by a simpler, more economic component with similar thermal-hydraulic characteristics
Design space		Region of possible (feasible) design conditions in a hyperplane of N dimensions (N being the number of design variables)
Theory of similarity		Theory used to infer the performance of a thermal system based on the performance of a different system that is geometrically and hydrodynamically similar
Hardware-in-the-loop	HIL	Technique used to test an electronic embebbed system by simulating the dynamic system associated with it
Reduced-scale test facilities		Facility where a smaller but geometrically and hydrodynamically similar system (similar to the actual system) is tested
Ejectors		Device wherein a high-pressure flow increases the pressure of another fluid at lower pressure by means of momentum and energy transfer

Glossary	Symbol/ Acronym	Definition
Venturi-based probe		Flow meter based on the venturi effect whereby an accelerating flow through a flow restriction (usually convergent divergent nozzle) experiences a static pressure drop
Thermal shocks		Mechanical stress caused by uneven thermal expansion in materials subjected to temperature changes
Thermal capacitance		Ratio of the heat added to an object to the resulting temperature change
Virtual turbine		Emulation of the compressor-turbine assembly in a fuel cell experimental facility
Startup combustor		Combustor used during initial operation (startup sequence) of the fuel cell
Servo-controlled valve		Electrically operated valve
Data acquisition system		System which converts analogue signals from physical measurements into digital data that can be managed by a computer
Transducers		Devices that converts a physical signal (for instance pressure) into an analogue electrical signal that can be managed by the data acquisition system
Surge test rig		Experimental facility to investigate the potential onset of surge in the operation of a compressor.
Variable speed electrical motor		Electric motor incorporating a controller that enables variable frequency (thus frequency or rotating speed)
Diaphragm		Membrane (sheet of semi-flexible material) used to separate gases – may also refer to diaphragm flow meters using these elements (volumetric meters)
Deceleration profiles		Pre-set sequence for turbomachinery shutdown manoeuvres
Resistive load bank		Set of resistances used to emulate the grid load in a fuel cell test rig

Glossary	Symbol/ Acronym	Definition
Transfer functions		Relation (usually linear and time-invariant) between the inputs and outputs of a subsystem in lumped-volume (zero-dimensional) modelling
Amplifiers		Electronic device that increases the power (intensity or voltage) of a signal
Bleed line		Extraction of pressurized gas from a turbine system (in gas turbine engines, from the compressor outlet flow)
Absorption chiller		Cooling system driven by a heat source and relying on a refrigerant (typically water) and liquid absorbant (typically lithium bromide) working in a two-phase cycle
Dryer		Device to remove water vapour from humid air
Superheater	SH	Heat adder that increases the temperature of vapour above the saturation temperature
Intercooler		Cooling system used to reduce the flow temperature between two subsequent compression stages
Lean premix combustor		Combustor where fuel and air are premixed (with an equivalence ratio lower than one) before combustion, in order to reduce the rate of NOx formation
Coriolis mass flow probe		Flow meter based on the Coriolis principle acting on a u-bent oscillating tube whose frequency and twist (deformation) depend on flow density and mass flow rate respectively
Transportation losses		Voltage drop along an electrical cable (typically applied to high voltage lines)
Cogeneration		Simultaneous production of heat and electricity
Reciprocating engines		Heat engine whose principle of operation relies on positive displacement enabled by linear motion (piston engines)
Trigeneration		Simultaneous production of heat, cooling power and electricity

Glossary	Symbol/ Acronym	Definition
Smart polygeneration grids		Decentralized electrical grid incorporating a number of operational procedures and appliances enabling much higher energy efficiency and security of supply
Hydro-methane	H_2/CH_4	Gas mixture composed of methane and a fraction (5% to 30%) of hydrogen
Hydrogen-rich fuels		Gaseous fuel with a high hydrogen content, usually coming from fuel reforming
Sabatier reaction		Endothermic chemical reaction of hydrogen and carbon dioxide at high temperatures to produce methane and water (also termed methanation reaction)
Net present value	NPV	Difference between the cash inflows and outflows of a project corrected to account for the time value of money
Internal rate of return	IRR	Discount rate for which the NPV is equal to zero
Payback period	PBP	Time needed to break even from the initial capital investment in a project
Levelized cost of electricity	LCE	Net present value of each kilowatt-hour produced over the lifetime of a power generation facility
Solar photovoltaic cells	PVs	Solid-state devices that convert solar energy into electricity directly
Run-of-the-river hydro power		Hydroelectric power plant with little or no water storage
Flywheels		Energy storage device in which energy is stored in the form of a rotating mass
Thermo-economic analysis		Combined economic and thermodynamic assessment of an energy system based on the First Law of Thermodynamics
Exergoeconomic analysis		Combined economic and thermodynamic assessment of an energy system based on the Second Law of Thermodynamics
Electrolyte cracking		Degradation of the electrolyte material in the form of small cracks
Microstructure coarsening		Change in a material microstructure leading to a change in material properties

Glossary	Symbol/ Acronym	Definition
High-speed turboalternators		Electric generator driven by turbomachinery rotating at high speed
Blended amines		Mixture of amines with different compositions used for carbon dioxide capture
Absorption systems		Carbon dioxide capture system based on the utilization of chemical sorbents (usually liquid)
Chemical looping		Fuel combustion based on two separate processes whereby the fuel does not react with pure oxygen in air but with an oxygen carrier (typically a metal oxide) acting as bed material in a fluidized bed reactor
CO_2 gas turbines		Gas turbine using carbon dioxide as the working fluid

Index

Hybrid Systems Based on Solid Oxide Fuel Cells: Modelling and Design, First Edition.
Mario L. Ferrari, Usman M. Damo, Ali Turan, and David Sánchez.
© 2017 John Wiley & Sons Ltd. Published 2017 by John Wiley & Sons Ltd.